Soil Erosion - Risk Modeling and Management

Edited by Shakeel Mahmood

Published in London, United Kingdom

Soil Erosion - Risk Modeling and Management
http://dx.doi.org/10.5772/intechopen.104013
Edited by Shakeel Mahmood

Contributors
Yaoxin Zhang, Yafei Jia, Keh-Chia Yeh, Chung-Ta Liao, Gabriel Nanfack, Moye Eric Kongso, Ashraf M. Khalifa, Hwat Bing So, Charles Galdies, Amy Zammit, Adam Gauci, Soheila Aghaei Dargiri, Davood Samsampour, Benaissa Kissi, Chafik Guemimi, Maria Dolores Rubio Cintas, Miguel Angel Parron Vera, Rachid Elkhayma, Shakeel Mahmood, Zara Tariq, Ian McIvor, Raphael Spiekermann, Thomas Mackay-Smith, John Mwangi, Charles K. Gachene, Stephen M. Mureithi, Boniface Kiteme, Partha Sarathi Mahato

First published in London, United Kingdom, 2024 by IntechOpen
IntechOpen is the global imprint of INTECHOPEN LIMITED, registered in England and Wales, registration number: 11086078, 5 Princes Gate Court, London, SW7 2QJ, United Kingdom

British Library Cataloguing-in-Publication Data
A catalogue record for this book is available from the British Library

Additional hard and PDF copies can be obtained from orders@intechopen.com

Soil Erosion - Risk Modeling and Management
Edited by Shakeel Mahmood
p. cm.
Print ISBN 978-1-80356-843-0
Online ISBN 978-1-80356-844-7
eBook (PDF) ISBN 978-1-83768-050-4

For EU product safety concerns:
IN TECH d.o.o., Prolaz Marije Krucifikse Kozulić 3, 51000 Rijeka, Croatia,
info@intechopen.com or visit our website at intechopen.com.

Meet the editor

Dr. Shakeel Mahmood is the chairperson of the Department of Geography, Government College University (GCU), Lahore, Pakistan. He has a distinguished academic background, including an MSc in Geography with a Gold Medal distinction, an MSc in geographic information systems, an MPhil in Applied Geomorphology, and a Ph.D. in Geography. Dr. Mahmood leads the Disaster Risk Modelling & Climate Change Research Group and actively participates in the GCU Press and the Disaster Risk Management Society. His significant contributions to the field have earned him recognition as an Integrated Research on Disaster Risk (IRDR) Young Scientist by the Chinese Academy of Sciences (CAS). With more than fifty research articles published in international journals, Dr. Mahmood is a respected author in the academic community. He also serves as a peer reviewer for several international journals and the academic editor of *Applied and Environmental Soil Science.*

Contents

Preface

Soil erosion is a pervasive environmental challenge with far-reaching implications for land, water, and ecosystem health. This book, *Soil Erosion – Risk Modeling and Management*, is a comprehensive exploration of the intricate science and practical strategies surrounding the understanding, prediction, and mitigation of soil erosion. Each chapter within this volume delves into a specific facet of soil erosion, offering unique insights, methodologies, and case studies worldwide.

Chapter 1 introduces the fundamental concepts of soil erosion-risk modeling and management, laying the groundwork for the subsequent chapters' in-depth examinations. In Chapter 2, the reader embarks on a journey to the Eastern Hindu Kush region in Pakistan, where the spatial quantification of soil erosion using the RUSLE approach reveals valuable insights into erosion dynamics. Chapter 3 shifts the focus to the Ol Bolossat catchment area, exploring the assessment of spatial soil loss with the aid of the Universal Soil Loss Equation (USLE) and geographic information systems (GIS). The application of the Analytic Hierarchical Process takes center stage in Chapter 4, as it explores the modeling of soil sensitivity to erosion in the Menoua Mountain watershed in West Cameroon. Chapter 5 focuses on a small watershed, providing a meticulous soil erosion risk analysis. Chapter 6 takes us to the downstream of a reservoir, shedding light on erosion control using in-stream weirs. The dynamics of coastal morphology near the Sundarban area in the Bay of Bengal are examined through remote sensing in Chapter 7. Chapter 8 explores the drivers and opportunities for woody vegetation and erosion management in pastoral hill country in New Zealand. Prediction of internal soil erosion in hydraulic works is the subject of Chapter 9, offering valuable insights into infrastructure stability. Lastly, Chapter 10 presents the utilization of rainfall simulators for designing and assessing post-mining erosional stability, offering innovative solutions for an increasingly relevant issue.

This book, compiled by experts in the field, serves as a valuable resource for researchers, policymakers, and practitioners engaged in the vital task of understanding, modeling, and managing soil erosion. It showcases diverse perspectives and methodologies, making it a comprehensive guide for addressing one of the most pressing environmental challenges of our time.

We hope this collection of chapters will inspire new research directions, inform sustainable land management practices, and contribute to preserving our precious natural resources.

Dr. Shakeel Mahmood
Chairperson,
Department of Geography,
Government College University Lahore,
Lahore, Pakistan

Chapter 1

Principles of Soil Erosion Risk Modeling

Soheila Aghaei Dargiri and Davood Samsampour

Abstract

It is anticipated that modern agriculture practice patterns will accelerate soil erosion in a negative way. Evaluating the long-term impact of various management strategies on a large farm is a gauge of the sustainable practices of soil nutrients. To find areas at risk, there are generally three different methods used: qualitative research, statistical approach, and model approach. Each of these approaches has distinctive features and applications. The use of geographic databases created using GIS technology has improved all techniques and strategies created recently. The sustainability of agricultural ecosystems worldwide is severely threatened by low or nonexistent attention given to environmental impact assessments, which also seriously threaten soil systems. Both conventional field-based methodologies and soil erosion modeling can be employed to quantify soil erosion. Agricultural automation has increased along with the accessibility of finer scale global level data, strengthening agri-environmental related modeling approaches. Due to the laborious, moment, limited flexibility, and noncomparability of field-based methods, soil erosion modeling has many advantages over these assessments. The examined models will be examined this season in the direction of wind erosion. The model is useful for forecasting and highlighting the areas most impacted by erosion while also saving time and resources.

Keywords: erosion risk assessment, modeling potential soil erosion, erosion hazard zones, erosion risk management, soil erosion types

1. Introduction

A serious issue is soil erosion, which averages 30–40 t/ha per year in South America, Africa, and Asia, and in the South Asia region is thought to be severe [1]. The agroecological efficacy in semiarid and arid regions is facing a significant impact from climate change, primarily due to an increased rate of land degradation [2, 3]. Due to the undulating to steep terrain and heavy rainfall, particularly in the first few years after establishment, soil erosion is typically higher in plantation farms. In order to maintain the productivity and fertility of the estates, appropriate soil conservation measures must be taken in order to reduce this soil erosion to a higher level. These measures included reducing soil erosion, strengthening the soil's structure to make it more resilient to detachment and transportation and more permeable to surface water, shielding the surface from the effects of rainfall, reducing runoff, and providing secure disposal options for excess runoff. Some of the features that have been seen include drainage systems, embankments, fences, cover crops, and stone terracing [4]. Despite the fact that soil is regarded as a mass containing nutrients, topsoil with nutrients has been drained in those fields over time, owing primarily to soil erosion. As a result, reliable and timely soil erosion monitoring in agricultural and plantation regions is crucial for developing soil preservation strategies and improving agricultural practices [5]. Numerous nations in the twentieth century experienced increased land loss as a result to raise human-induced soil erosion [6]. The most fertile topsoil can be lost due to erosion, which lowers soil productivity. Investigating soil loss mechanisms and determining the risk of soil erosion are crucial for planning future management of soils, preservation, and land-use activities [7–9].

Soil erosion can be evaluated using traditional field-based techniques and soil erosion modeling [10]. Agricultural computerization has increased along with the accessibility of finer-scale global-level data, which has improved the potency of modeling techniques associated with agriculture and the environment [11]. Field-based approaches to measuring soil erosion are labor-intensive, time-consuming, limited in flexibility, and incomparable, whereas soil erosion modeling has numerous benefits over these approaches [12, 13]. Several methods for modeling soil erosion have been established in recent years with varying needs for input and complexity [5]. Applications, specifications, intended uses, and the type of data each model provides vary significantly [14]. Soil erosion modeling, which is employed in place of traditional methodologies, is the most practical and trustworthy instrument for evaluating soil erosion and enabling the appropriate selection of soil erosion management strategies [12]. One of the primary reasons for the widespread use of soil erosion modeling around the world is unquestionably its high degree of adaptability and data accessibility, as well as its sparse parameterization, broad research, and comparability of results, which allow the model to be applied to almost any situation or geographical area [13]. The Universal Soil Loss Equation (USLE) and its associated models, which are widely employed to address soil erosion, are among the most extensively used designs. These soil erosion algorithms have been applied in a variety of contexts worldwide, making them well-known [15]. In this chapter of the book, soil erosion, risk modeling, and management using the model are discussed.

2. Soil erosion

2.1 Soil erosion definition

"The dirt on top of a field is naturally eroded by forces such as wind and water. This occurrence is known as soil erosion" [16]. It leads to a constant loss of topsoil, ecological deterioration, soil collapse, etc. Soil erosion is a continuous process that can occur either slowly or quickly. Soil erosion (**Figure 1**) is caused by the loosening or washing away of dirt particles in ravines, seas, waterways, or distant lands [9].

2.2 Types of soil erosion

The erosive process' rate or the agent accountable for it is used to categorize objects. Fast or gradual and natural or human-caused soil erosion can all be categorized. Water currents and windstorms are the main causes of soil erosion caused by natural processes, but human activity can also make the problem worse [17].

Figure 1.
Soil erosion (@Frederick J. Weyerhaeuser).

In **Figure 2**, types of soil erosion are drawn with slight modifications (by HARM VENHUIZEN Associated Press/Reporting for America).

2.2.1 Water erosion

This sort of soil erosion, as its name implies, is brought on by water and denotes the removal of topsoil as a result of precipitation, snowmelt, floods, or improper irrigation.

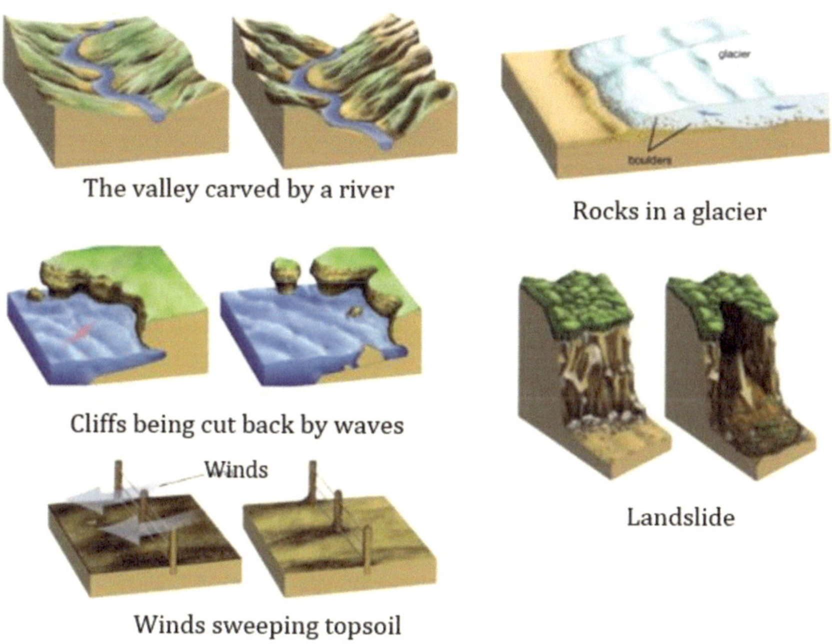

Figure 2.
Type of soil erosion.

As a result, it can happen as a result of farming activity or extreme weather. Water is more destructive on bare land and during periods of heavy rain or melting [18].

2.2.2 Wind erosion

Another reason for erosion is dust storms, which have become more frequent in recent years, especially in arid areas. When the earth is level, acceptable, and dry, erodibility increases; nevertheless, hills reduce wind force and make it more difficult to remove rough and heavy particles [19].

2.2.3 Anthropogenic soil erosion

This typically happens as a result of anthropogenic forces, and both direct and indirect human activity can cause soil erosion. For instance, mining and quarrying have an immediate impact. The topsoil is impacted by unsustainable management in indirect ways, which aggravates agricultural and forest standing erodibility [20].

2.3 The risks of soil erosion

More than just the loss of fertile land is a consequence of soil erosion. It has resulted in increased sedimentation and pollution in streams and rivers, blocking these waterways and resulting in a fall in fish populations and other species. Degraded areas are also frequently less able to retain water, which can make floods worse [21].

2.4 Mechanisms for erosion

Detachment (from the ground), movement (by water or wind), and deposition (frequently in areas where we do not want the soil such as streams, lakes, reservoirs, or deltas) are the three processes involved in erosion [22].

2.5 Reason for soil erosion

The following factors are significant contributors to soil erosion:

2.5.1 Rainfall and flooding

Four different types of soil erosion are brought on by rainstorms of greater intensity: Sheet and rill erosion, sheet erosion, gully erosion, and splash erosion. In locations with extremely heavy and frequent rainfall, a sizable amount of soil is lost because rainfall drops scatter the soil, which is then washed away into the nearby rivers and streams. Running water during floods also obliterates a tremendous amount of soil by creating holes, rock-cut basins, etc. [16].

2.5.2 Agriculture

Soil erosion is mostly caused by agricultural practices. The ground is disturbed by agriculture. To plant new seeds, the trees are cut down, and the ground is plowed. The land is left fallow during the winter because the majority of the crops are grown in the spring. During the winter, the earth erodes the most. Tractor tires also leave grooves in the ground that operate as a natural waterway. Wind erosion removes fine soil particles [16].

2.5.3 Grazing

Grazing animals consume the grasses on the land and clear it of its flora. The dirt is disturbed by their hooves. Additionally, they remove plants from the roots. As a result, the soil becomes more permeable to erosion [23].

2.5.4 Logging and mining

The logging process necessitates the removal of multiple trees. Trees firmly hold the soil in place. The forest cover protects the soil from intense rain. The leaf litter that protects the soil from erosion is also removed during logging. Additionally, mining activities harm the ecology and increase soil erosion risk [24].

2.5.5 Construction

The soil is at risk of erosion due to the building of structures and roadways. The trees and meadows are destroyed for development purposes, exposing the soil, and making it vulnerable to erosion [25].

2.5.6 Rivers and streams

The dirt particles are carried away by the moving waters of streams and rivers, causing a V-shaped erosion action [26].

2.6 Heavy winds

Small soil particles are carried away by the wind to distant countries when the weather is dry or in semiarid regions. During dry periods or in semiarid regions, the wind carries minute soil particles to far-off nations [24].

2.7 Effects of soil erosion

Erosive processes have an effect on agricultural productivity, deteriorating rural communities' living conditions and well-being (both for individual farmers and agricultural cooperatives). Farmlands that have been eroded over time lose their soil fertility, deteriorate, and are no longer suited for farming. In addition, erosive activities severely harm the environment, decreasing biodiversity and disrupting the balance of ecosystems. However, soil erosion is an issue for other reasons as well. Oftentimes, it goes unnoticed, causing irreversible land decay [27].

The principal consequences of soil erosion include:

2.7.1 Loss of arable land

Crop production is not supported, and agricultural output is reduced since soil erosion damages the top fertile layer of the soil, which is rich in the nutrients needed by plants and the soil [28].

2.7.2 Obstruction of waterways

In addition to fertilizers and other chemicals, agricultural soil also contains pesticides and insecticides. Thus, the waterways where the soil flows get contaminated. Flooding results from sediments building up in the water and increasing water levels [29].

2.7.3 Polluting the air

The mixing of dust particles in the atmosphere is the main source of air pollution. When inhaled, some toxic compounds, such as petroleum and insecticides, can be quite dangerous. When the winds blow, the arid and semiarid regions' dust plumes generate extensive pollution [30].

2.7.4 Desertification

Soil erosion is a primary contributor to desertification, transforming habitable areas into deserts. Deforestation and damaging land use exacerbate the situation, resulting in biodiversity loss, soil degradation, and ecosystem changes [31].

2.7.5 Infrastructure destruction

Soil silt deposition in dams and along their banks might limit their efficiency. As a result, it has an impact on infrastructure projects such as dams, embankments, and drainage [24].

2.7.6 Losses of topsoil

The removal of topsoil by water or wind has a significant negative impact on field fertility since it is the layer of the ground that contains the most organic matter and nutrients. This is why soil erosion on agricultural land is so important. Additionally, rills or gullies make it very difficult to cultivate eroded fields [32].

2.7.7 Soil acidification

The agricultural ground may become more acidic due to a lack of organic matter, delaying crop growth, and exposing it to water and wind [33].

2.7.8 Losses in planting material

Due to agricultural losses and decreased farmer profitability, water streams and dust storms kill seedlings and remove seeds from the fields [34].

2.7.9 Water contaminant

Other implications of soil erosion include sedimentation and the deterioration of irrigation water quality due to chemical pollution of water bodies from the crops [35].

2.8 Soil erosion affects the environment

There are other problems beside the detrimental consequences on agriculture. Plants and aquatic life suffer due to soil erosion, which also results in biodiversity loss, sedimentation, and frequent flooding [36].

2.8.1 Events of regular flooding

Due to the stabilizing effects of tree roots, transforming forests into pastures or fields increases the likelihood of flooding and waterlogging since these areas lose their capacity to allow for infiltration [37].

2.8.2 Clogged waterways and polluted aquatics

In addition to causing sedimentation in areas with higher altitudes, eroded particles suffocate water pumps, dams, and grass-lined streams. Frequently, chemicals that are harmful to people, animals, and aquatic life are present in the water currents from fields [38].

2.8.3 Loss of biodiversity

Because many creatures are robbed of their native homes, eroded lands have scarce vegetation and eventually become completely naked. The loss of biodiversity causes environmental imbalance [39].

2.8.4 Reduced greenhouse gases sequestration

The loss of carbon-sequestering plants on our world might be slowed or stopped with sustainable management, lowering greenhouse gas emissions caused by deforestation. Trees and vegetation are excellent at storing carbon dioxide, but they can rarely develop on degraded terrain. Additionally, soils themselves can function as CO2 sinks. According to Aberdeen University Professor Peter Smith, the planet can store almost 5% of man-made greenhouse gases per year [40].

2.9 Solutions to soil erosion

The kind of soil, terrain, local climate, and appropriate agricultural practices such as crop rotation or tillage methods all play a role in the decision-making process for controlling soil erosion [41]. Analyzing the efficacy of implemented techniques and customizing them for different fields is crucial [42]. The key to success is early problem discovery and the selection of appropriate solutions based on the severity of the issue. In the early stages of soil erosion, for instance, replanting, growing cover crops, or mulching can be successful treatments because vegetation protects fields from harm brought on by water runoffs, rainfall, and wind. In severe cases, the effects can be lessened by terrace farming or check dams [43]. In order to stabilize the ground and slow down water streams, further soil erosion management techniques include contour cropping and perennial plantings with robust root systems [44].

2.10 Soil erosion prevention

The major environmental problem of soil erosion. It is necessary to take action to solve this issue. Some strategies for preventing soil erosion include the ones listed below:

1. To stop soil erosion, add mulch and rocks to the grass and plants below.

2. Plant trees to prevent soil erosion on bare land.

3. Mulch sheeting may be utilized to stop slopes from eroding.

4. Place a number of fiber logs to prevent any soil or water from washing away.

5. Create a wall at the bottom of the slope to aid in preventing soil erosion.

6. Proper drainage should be installed in every home so that water can be collected in the right places [45].

2.11 Key points of soil erosion

The topsoil is naturally being worn away, but human activity has sped up the process. It usually happens as a result of clearing vegetation or taking any other activity that dries out the ground. Soil erosion is a problem that can be caused by farming, grazing, mining, building, and recreational activities. One effect of soil erosion is land degradation, which has led to a rapid increase in river pollutants and sedimentation, choking up water bodies, and reducing the diversity of aquatic life. Floods occur when deteriorated areas lose their capacity to hold water. Although there are numerous obstacles to overcome when trying to stop soil erosion, there are also solutions to it. Farmers and the community that depends on agriculture for food and employment place the greatest importance on the health of the soil [46].

3. Soil erosion models

Several theoretical frameworks emphasize the significance of protracted (whether natural or geological) erosion in shaping the topography. A multitude of erosion models has been devised to gauge the potential outcomes of expedited soil erosion or soil erosion stemming from anthropogenic actions [47]. The phenomenon of tillage erosion is frequently neglected by models, and the development of soil erosion models is more commonly observed in agricultural landscapes rather than in naturally vegetated areas such as forests or rangelands. The preponderance of erosion models concentrates solely on the phenomenon of soil erosion due to water, whereas some other models center their attention exclusively on the issue of erosion at mining locations [48]. The fundamental aim of the majority of soil erosion models is to predict customary levels of soil depletion (frequently an annual mean rate) within a given area *via* the utilization of diverse land management practices such as a plot, a field, or a catchment/watershed [49]. Some models used to predict erosion are founded on statistical principles, whereas other models rely on mechanical or physical principles [50]. The Water Erosion Prediction Project erosion model (WEPP) and the Revised Universal Soil Loss Equation (RUSLE) represent two of the most widely utilized soil erosion models in North America [51]. The majority of mine land erosion research focuses on setting up or improving RUSLE parameters. Gully erosion is frequently excluded from soil erosion models because it is challenging to forecast these essential erosional features [52].

3.1 Principles of erosion modeling

The complex interactions among a number of elements of the Earth's system such as precipitation, surface, subsurface, groundwater movement, vegetation growth, soil detachment, transport, and deposition, which lead to patterns of erosion and deposition in both time and space [53]. This section concentrates on the management of rainfall, runoff, vegetation cover, and soil characteristics as ingredients without going into great detail about the methodology used to calculate the values of these elements [54]. Here, an extensive mathematical representation of erosion and sediment transport processes is presented, and an association between models with various levels of complexity and experience is derived, showing the common principles [55].

3.1.1 Soil Erodibility models by type

Three types of soil erosion models can be distinguished: theoretical, physics-based, and empirical or statistical models. These models can be divided into three categories based on the physical processes they replicate, the model algorithms that represent these processes, and how much they rely on data [14]. The Universal Soil Loss Equation (USLE), along with its derivatives, the Revised Universal Soil Loss Equation (RUSLE), and the Modified Universal Soil Loss Equation (MUSLE) represent prominent instances of empirical models. They continued by saying that because they may be used in scenarios with sparse data and parameter inputs, empirical models are the most basic models. Additionally, they are particularly helpful in locating the sources of sediment and nutrient production [56].

3.1.2 Reduced erosion process models

It is possible to create equations for streamlined models of erosion processes that satisfy easy-to-compute models with freely available data, interactions between two restrictive conditions, rainfall, runoff, the local land cover's condition, erosion, and sediment movement [57].

- Limited separation capacity.
- Sediment transfer capacity is limited.

3.1.3 Estimating erosion rates: Techniques

The classification of soil erosion assessment approaches is shown in **Figure 3**. Many different scientific techniques and modeling strategies have been used to build a number of soil erosion models. There exist three distinct categories of soil erosion models, namely physics-based, empirical, and conceptual models. These models are dependent on the underlying algorithms utilized, with each model type tailored to the specific nature of the algorithms employed [58, 59].

- Models that use only statistics
- A comprehensive analysis has been conducted by professionals on a compilation of data regarding erosion rates across both naturally occurring and reclaimed locations situated in close proximity to both natural and man-made ecosystems.
- Examinations of prevailing erosive or depositional structures to determine the mean erosion rates are imperative. The age of these attributes should be duly

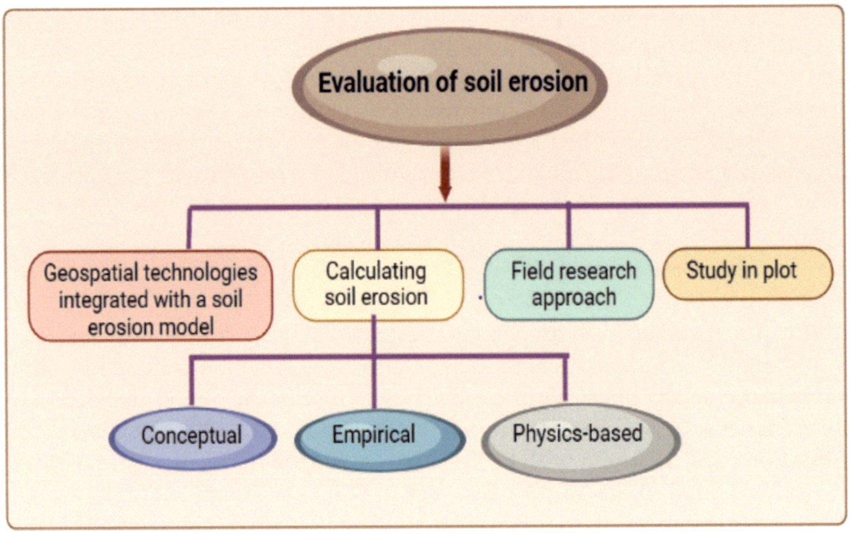

Figure 3.
Methods for classifying soil erosion assessments.

ascertained (or determined by dating deposits). The analysis of antiquated aerial photography is a common practice.

- Empirical models that are site-specific that link slope, watershed size, and rainfall
- Observations of erosion in flumes under replicated flow or rainfall situations
- Models of gully erosion with a physical basis
- Statistical or erosion mechanics-based landform and landscape scale models that are typically based on GIS are used to forecast changes in topography and erosion rates.
- Using data from watershed monitoring and sediment budget models [60].

3.1.4 The method for modeling erosion risk

An erosion risk modeling approach combined with routine field research can produce reliable decision support beneficial for the effective management of soil erosion risk [61–63]. A global initiative to predict soil loss has been the development of empirical and process-based models [14, 64–66]. The majority of the effort has gone toward thoroughly assessing the risk of soil erosion [67–69]. Habib-Ur-Rehman [70] employed a methodology centered on the process to prognosticate soil erosion on a regional scale. Numerous models of soil erosion and sediment transportation have been created on a global scale to determine the rates of sediment and nutrient movement in various land use systems. These models can be classified into three distinct groups, namely conceptual models, physical models, and empirical models, as stated by Merritt et al. [14]. The GIS-based models, namely the USLE, WEPP, AGNPS, LISEM, and EUROSEM, exhibit considerable dissimilarities in terms of their complexity, inputs and prerequisites, methodologies, visual representations, intended application domains, and output data formats [14, 71].

3.1.5 Modeling erosion using GIS

The application of GIS in erosion models enables the storing of georeferenced data, computation of input parameters for multiple scenarios, geographical analysis of modeling outputs, and effective display. GIS is utilized for the statistical analysis and modeling of erosion processes found in remote sensing data. In the early 1990s, the Geographic Resource Analysis and Support System (GRASS) provided an environment for creative work on the integration of GIS with hydrologic and erosion modeling [72, 73]. The primary use of geospatial erosion models is found in the fields of agriculture, soil conservation, minimization of silt contamination, and sustainable military management as evidenced by various sources [64, 74]. One of the initial hillslope erosion models utilized in GIS was the Universal Computing Soil Loss Equation (USLE), which was applied to analyze the impact of recent wildfires on forests and hillslopes [75]. Furthermore, topographic parameters derived from digital elevation models (DEMs) are extracted to support the analysis. Moore and Wilson come next, then Moore and Birch [73, 76], set the stage for USLEAP-Applications to Landscapes with Complex Topography and the connection between unit flow power theory and USLE. The popularity of this strategy has led to multiple USLE installations supported by GIS for challenging topographic settings [77, 78]. Among the more recent GIS applications of USLE coverage area size, large watersheds with mapped land cover from remote sensing images are just a few examples [79–82].

3.1.5.1 Making erosion estimation easy with GIS tools

The RUSLE model was originally formulated with the intent of assessing the potential for soil erosion in small and localized watersheds. Nevertheless, due to the wide distribution, rapidity, and concerns related to the quality of water, the utilization of the RUSLE framework inherently presents drawbacks in relation to expenses of implementation, adequacy of site representation, and precision of anticipated outcomes [76, 83]. The spatial distribution of soil erosion using the traditional RUSLE model is often challenging to map, thus posing a considerable difficulty [83]. The proliferation of GIS-based models at the regional level has surged significantly subsequent to the innovation of GIS technology. Various researchers have reported that the utilization of GIS technology in tandem with erosion models, such as the RUSLE, has considerably boosted the efficacy of assessing the spatial dissemination and magnitude of erosion hazards, while simultaneously reducing costs and augmenting precision. These findings have been documented in the scholarly literature [77, 84–88].

3.1.6 Soil erosion using the RUSLE model

Satellite-based remote sensing and the utilization of geographic information systems (GIS) are indispensable instruments in the evaluation of soil erosion in spatial contexts. This is due to their remarkable ability to extract, identify, and modify land features, as well as their seamless integration with the Revised Universal

Soil Loss Equation (RUSLE) [83, 89–91]. The RUSLE model exhibits extensive usage and has an extensive record of validation. It is noteworthy that its limitations have been thoroughly established [92, 93]. African soil erosion rates have been predicted and assessed in a number of studies using RUSLE, with a focus on highlands and river basins [94–96]. Nevertheless, there is a dearth of research on identifying possible erosion and simulating the danger of soil erosion in built-up metropolitan settings [97–99]. In contemporary times, noteworthy progressions in urban planning and the mitigation of soil erosion have demonstrated that land managers and policymakers hold a greater degree of significance toward the spatial distribution of soil erosion risk as opposed to the factual values for soil loss [100].

3.1.6.1 Using RUSLE to estimate parameters for soil erosion risk

The preeminent and uncomplicated digital manifestation of the Universal Soil Loss Equation (USLE) was formulated with the aim of computing the yearly soil erosion per unit region predicated on erosion attributes [101, 102]. The RUSLE model is regularly employed to forecast the mean yearly soil depletions due to sheet and rill disintegration along with exhibiting the geographic arrangement of potential erosion hazard [51, 97, 102–106]. The assessment of soil erosion risk through the utilization of the RUSLE model involves the consideration of several critical factors, including the slope length and steepness factor (LS), the land cover and management component (C), the support practice factor (P), and the rainfall erosivity factor (R) [101]. In the current investigation, the peril of soil erosion was spatially allocated, and conceivable erosion was charted by using C and P factors as identifying elements (C and P = 1).

According to reference [101], the Revised Universal Soil Loss Equation (RUSLE) stipulates that:

$$\mathrm{A} = \mathrm{R} * \mathrm{k} * LS * \mathrm{C} * \mathrm{P}$$

Figure 4 presents an exemplification of the procedure through which the model's input parameters are procured from diverse sources that comprise rainfall data, soil attributes that include soil texture, hydraulic conductivity, and organic matter content, as well as topographical characteristics such as slope length and percentage. These attributes are acquired from elevation digital models (DEMs) and satellite imagery. The RUSLE model encompasses several inputs such as topography (LS factor), crop cover (C factor), soil erodibility (K factor), rainfall erodibility (R factor), and soil erodibility (K factor), among others [107].

3.1.6.2 Techniques for evaluating RUSLE factors in a GIS environment

The application of the Revised Universal Soil Loss Equation (RUSLE) has been extensively implemented in various settings, including tropical watersheds with mountainous terrain, expansive watersheds, those dominated by agricultural practices, locales exhibiting discernible wet and dry seasons, and areas undergoing dynamic transformations in terms of land coverage patterns, agricultural farmland utilization, and developmental activities. The RUSLE model comprises three primary databases, namely the climatic and survey database, the crop database, and the soil data. The climatic and survey database contains monthly temperature

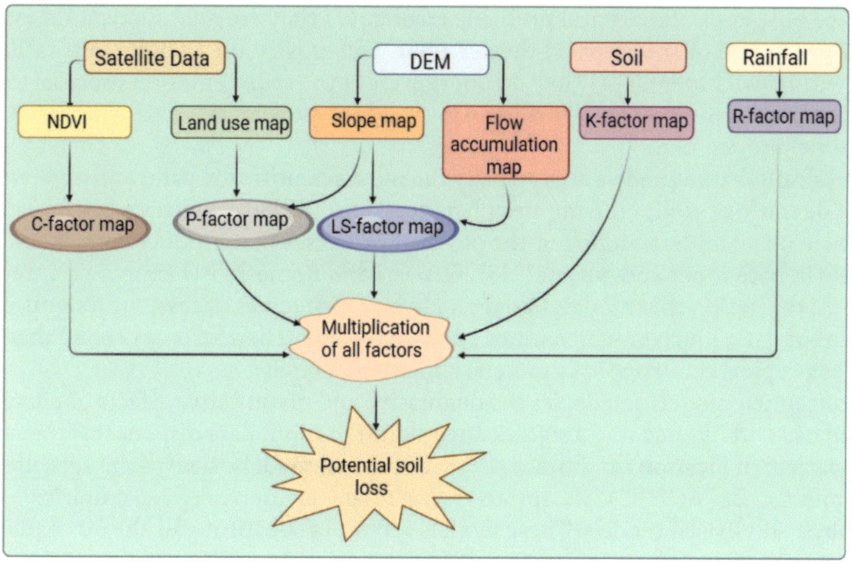

Figure 4.
Flowchart for RUSLE-based estimation of soil erosion.

and precipitation data, as well as contours essential for the computation of factors such as erosivity, slope length, and steepness (LS). On the other hand, the crop database contains crucial data required for the determination of the surface cover factor (C). Lastly, the soil data includes relevant information on soil survey and soil characterization. The RUSLE model incorporates the five variables enumerated in eq. 1 to compute the mean annual soil erosion loss [108]. Estimation of the various components of the model, which is rooted in a significant corpus of research, is a prevalent approach to employing the Revised Universal Soil Loss Equation (RUSLE). Prior scholars have employed diverse techniques to compute these variables such as utilizing meteorological data, geological and soil maps, satellite imagery obtained remotely, empirical formulas, and digital elevation models (DEM) sourced from multiple origins [109].

3.1.7 Fundamental issues with erosion modeling

The inherent complexity of landscape systems, regional variation, and a lack of data makes distributed erosion models difficult [14]. A novel investigation is imperative for soil erosion prediction due to the fact that, despite the extensive efforts devoted to soil erosion assessment at the plot or catchment level, the quantitative estimation of soil erosion, which is regionally distributed, has not been comprehensively tackled [110]. The basic difficulty with erosion risk models is validation due to a lack of data to compare model projections with actual soil loss [111]. The data sources from which empirical models were constructed limit their application to locales and ecological circumstances [14]. Smith [112] claims that empirical models are particularly useful in a number of situations since they are the only ones that can be utilized when there is little available data. They have the following restrictions, among others; They have several drawbacks, including the following: (1) they are based on statistical analysis of significant factors in the soil erosion process and produce only approximate and probable results; (2) they are not practical for event-based prediction of soil loss; (3) they estimate soil erosion on a single slope rather than within catchments; (4) they do not represent the sedimentation process; (5) they are limited to sheet and/or rill erosion; and (6) they merely take changes in soil over time into account.

Physically-based models are typically the most scientifically valid and applicable to a wide range of soils, climatic circumstances, and land use scenarios because they are based on an understanding of the physical processes that produce erosion and are flexible in both input and output [113]. Ganasri and Ramesh [81] expressed agreement on the notion that models based on physical principles necessitate a significant amount of data, much of which is not easily accessible. They further posited that this implies a typical challenge in parameterizing such models.

Conceptual models can depict the qualitative and quantitative effects of changes in land use without needing a significant amount of input data dispersed across a wide variety of locations and times [14]. Conceptual models, such as the agricultural nonpoint source (AGNPS), occupy an intermediate position between empirical and physically-based models. These models serve as a substitute for the mechanical components of the system in question [56].

3.1.8 Soil erosion: risk management

When surface vegetation is removed or physically disturbed, the soil is more susceptible to erosion. Seasonally extremely dry circumstances raise the danger of erosion, where there is less vegetation due to inadequate crop and pasture growth [34].

Important management techniques that have an impact on the risk of soil erosion:

1. How often, how intensely, and when tillage activities occur.

2. the amount and kind of surface cover [114].

The greatest risks related to feeding happen in late summer and autumn if the supply of feed and the amount of cover of each year's crop and grass leftovers is decreasing. The majority of the erosion risk is caused by showing practices such as cultivation and burning of stubble. Grazing management is additionally a significant factor, in particular during dry years, especially if there are more than two subsequent dry seasons [115]. The safeguarding of soil from erosion has significantly increased with the implementation of more environmentally friendly land management techniques such as no-till sowing and stubble retention. In order to avoid soil disturbance and maximize residue protection on the soil surface, no-till sowing entails placing the seed in a small opening in the soil [116]. Water-repellent soils increase the danger of soil erosion by causing poor plant germination and limited development of plants. A key tool for preventing soil erosion is the spreading, digging, and spading of clay to cultivate water-repellent soil [117]. These methods of soil alteration are commonly employed in the Southern Mallee, Upper South East, and portions of the Eyre Peninsula, where there are vast expanses of naturally water-repellent soils [118]. The utilization of these methodologies engenders an escalation in the clay concentration of the topsoil, thereby fortifying and safeguarding the soil from erosion [119]. Livestock can be withdrawn from fields using confined feed before the ground covering deteriorates too far. It is a crucial method for stopping erosion during dry seasons, as well as in the late summer and early fall when vegetation is dwindling [120].

3.1.9 Risk assessment of soil erosion

Despite the potential cost of collecting field data for risk assessments at elevated levels, it remains a crucial factor in the formulation of efficacious policies and strategies concerning the conservation of soil and water resources [121]. The technological progress in geographic information systems (GIS) and satellite imagery provides practical avenues for surveying, classifying, identifying, and tracking land use and soil at diverse levels. In numerous research endeavors, models aimed at evaluating soil erosion have been established and executed through the utilization of satellite data and geographic information system (GIS). Remarkably, it has been disclosed that these models are at times more efficacious and accurate in detecting and linking the peril of soil erosion compared to field survey data. This, in turn, confers invaluable insights for resource management and soil conservation planning [122]. Overall, the assessment of soil erosion employs both qualitative and quantitative approaches. Despite the availability of measures for approximating soil erosion volumes and rates, the determination of the severity of the risk is conducted qualitatively. The methodology utilized in qualitative evaluation encompasses a broad spectrum of techniques such as picture categorization [123], index linking [124], photo analysis [125], field research statistics [126], and photo interpretation. The soil erosion risk map is produced through the utilization of the index coupling procedure, which is a qualitative evaluation technique that utilizes remote sensing images and GIS [127]. The aforementioned methodology has been demonstrated to be a plausible and economically feasible approach for gauging the likelihood of erosion [128, 129]. Several quantitative techniques, such as USLE, RUSLE, CORINE, PESERA, and WEPP, employ methodologies for constructing models [130–133]. The determination of the average annual soil loss per unit area of soil over a prolonged period can be ascertained through the utilization of the Universal Soil Loss Equation (USLE), a widely employed and uncomplicated soil erosion model [134]. Additionally, the Revised Universal Soil Loss Equation (RUSLE) has been established as a novel technique that integrates current data, which supersedes the USLE approach [92]. There exists alternative, less commonly utilized methodologies for appraising soil erosion depletion, including the collaboration of data on the environment (CORINE) model, which was derived from a USLE model for assessing the erosion vulnerabilities and attributes within the member states of the European Union (EU) [135]. The PESERA model, designed to predict long-term average erosion rates at a 1 km resolution, has been predominantly adopted by a large portion of Europe [136]. The water erosion prediction project (WEPP) computational model is an uninterrupted, simulation-based, and distributed parameter framework for soil erosion prediction that is equivalent in status to the USLE and RUSLE models [137]. However, the authentication of qualitative erosion models presents obstacles due to the requirement for extensive proof, fresh resources, and the preparation of qualified personnel [138].

3.1.10 Global soil erosion analysis studies

Water-induced soil erosion is a significant factor in the global degradation of land [139–142]. The nutritious topsoil, which contains the majority of organic matter and nutrients, is lost due to erosion [143, 144]. According to the global evaluation of land degradation (GLADA) conducted by the United Nations Environment Program (UNEP), it has been observed that there has been a degradation of 1.1 billion hectares of land globally due to soil erosion. The issue of land degradation has been most prevalent in Asia, accounting for a significant portion of the total at 48%, followed by Africa at 21%, Latin America and the Caribbean at 15%, Europe at 10%, Oceania at 8%, and North America at 5% [145]. In areas prone to drought, it was estimated that the yield loss of agricultural produce was associated with a significant decrease of 55% [146]. The regions exhibiting the greatest levels of soil erosion were observed in Africa and Asia, predominantly due to the presence of intense precipitation that is highly erosive in nature [120, 122], larger population [123], and despite the loss of natural vegetation, there has been a proportional increase in agricultural land and urban areas [124, 125]. Researchers have shown that soil erosion rates vary widely among countries, continents, and climate zones [28, 112, 118–121]. In conjunction with these alterations, a significant reduction in biodiversity and ecological services has been observed [147–149]. According to a recent scholarly investigation [113], it has been anticipated that there will be an escalation in soil erosion until the water on a global scale by 30–66% by the year 2070 due to the projected alterations in climatic conditions and land use. This phenomenon is expected to have a more pronounced impact on the global poor. This necessitates additional research to enhance our knowledge of the key factors influencing soil erosion throughout a broader variety of geographical areas [142]. This necessitates additional research to enhance our knowledge of the key variables influencing soil erosion over a larger variety of geographical regions. Additional knowledge on land cover and its management practices is of particular importance because it will help with planning, implementing, and evaluating mitigation efforts [150].

3.1.11 Limitations of the model

Recent research calls for increased field-based erosion assessment and monitoring because it is challenging to anticipate erosion potential using models [151]. In addition to being instruments for simulation and predictability, models are tools for understanding. As a result, models can be thought of as both instruments for understanding and tools for controlling variables, trends, and disparities between data that are distinct in terms of space or time [152]. There is always a potential that what models transmit is insufficient or even incorrect, and these deficiencies may not be obvious to the model user or the recipient of the model output as models are also used to communicate research and its results [153].

4. Conclusion and future directions

Soil erosion modeling is developing due to the availability of highly precise geographical and meteorological data for monitoring intra-annual fluctuations in temperature, vegetation, and management practices. Accurate soil erosion modeling is required to make informed decisions about planning, management, and policy at both small and large spatial scales. Based on a comprehensive review of the literature, it has been observed that the RUSLE model has gained widespread usage and has been proven to be an effective tool in estimating the amount of soil loss caused by erosion across various regions of the globe. The RUSLE model has proven to be a valuable tool when implemented on a local scale. However, the integration of RUSLE with GIS methodologies has greatly improved the assessment of soil erosion that is geographically distributed over large catchment areas. Based on an analysis of the existing literature, it has been determined that the fundamental components of the model can be derived from various sources of data, including digital elevation models (DEMs), meteorological data, cartographic depictions of soil, and remote sensing imagery. GIS technology facilitates the examination of a broad study region, specifically in the context of soil erosion studies, by virtue of its requisite capabilities. Nevertheless, the identification of an appropriate model that can accurately assess soil loss from all forms of erosion at a given site remains a challenging endeavor.

Author details

Soheila Aghaei Dargiri* and Davood Samsampour
Faculty of Agriculture and Natural Resources, Department of Horticultural Sciences, University of Hormozgan, Bandar Abbas, Iran

*Address all correspondence to: s.aghaei.phd@hormozgan.ac.ir

References

[1] Iqbal MN, Oweis TY, Ashraf M, Hussain B, Majid A. Impact of land-use practices on sediment yield in the Dhrabi watershed of Pakistan. Journal of Environmental Science and Engineering A. 2012;**1**(3A):406-420

[2] Alam S, Fatima A, Butt MS. Sustainable development in Pakistan in the context of energy consumption demand and environmental degradation. Journal of Asian Economics. 2007;**18**(5):825-837

[3] Ahmad F. Land degradation pattern using geo-information technology for Kot Addu, Punjab province, Pakistan. Global Journal of Human-Social Science Research. 2013;**13**(1-B):1-16

[4] Bandara JS, Chisholm A, Ekanayake A, Jayasuriya S. Environmental cost of soil erosion in Sri Lanka: Tax/subsidy policy options. Environmental Modelling & Software. 2001;**16**(6):497-508

[5] Piyathilake I, Udayakumara E, Gunatilake SK. GIS and RS based soil erosion modelling in Sri Lanka: A review. Journal of Agricultural Sciences – Sri Lanka. 2021;**16**(1):143-162

[6] UNDP BS. Power, Poverty and the Global Water Crisis. New York: Human Development Report, United Nations Development Programme; 2006. p. 440

[7] Parveen R, Kumar U. Integrated Approach of Universal Soil Loss Equation (USLE) and Geographical Information System (GIS) for Soil Loss Risk Assessment in Upper South Koel Basin. Journal of Geographic Information System. Jharkhand. 2012;**4**(6):588-596

[8] Ashiagbor G, Forkuo EK, Laari P, Aabeyir R. Modeling soil erosion using RUSLE and GIS tools. International Journal of Remote Sensing. 2013;**2**(4):1-17

[9] Serpa D, Nunes J, Santos J, Sampaio E, Jacinto R, Veiga S, et al. Impacts of climate and land use changes on the hydrological and erosion processes of two contrasting Mediterranean catchments. Science of the Total Environment. 2015;**538**:64-77

[10] Zachar D. Chapter 3 problems and methods of soil erosion research. Developments in Soil Science. 2008;**10**:137-203

[11] Wimalasiri EM, Jahanshiri E, Chimonyo VG, Kuruppuarachchi N, Suhairi T, Azam-Ali SN, et al. A framework for the development of hemp (Cannabis sativa L.) as a crop for the future in tropical environments. Industrial Crops and Products. 2021;**172**:113999

[12] Moehansyah H, Maheshwari BL, Armstrong J. Field evaluation of selected soil erosion models for catchment management in Indonesia. Biosystems Engineering. 2004;**88**(4):491-506

[13] Alewell C, Borrelli P, Meusburger K, Panagos P. Using the USLE: Chances, challenges and limitations of soil erosion modelling. International Soil and Water Conservation Research. 2019;**7**(3):203-225

[14] Merritt WS, Letcher RA, Jakeman AJ. A review of erosion and sediment transport models. Environmental Modelling & Software. 2003;**18**(8-9):761-799

[15] Igwe P, Onuigbo A, Chinedu O, Ezeaku I, Muoneke M. Soil erosion:

A review of models and applications. International Journal of Advanced Engineering Research and Science. 2017;**4**(12):237341

[16] Zachar D. Soil erosion. New York: Scientific Publishing Company; 2011

[17] Haregeweyn N, Tsunekawa A, Nyssen J, Poesen J, Tsubo M, Tsegaye Meshesha D, et al. Soil erosion and conservation in Ethiopia: A review. Progress in Physical Geography. 2015;**39**(6):750-774

[18] Vrieling A. Satellite remote sensing for water erosion assessment: A review. Catena. 2006;**65**(1):2-18

[19] Zobeck TM, Van Pelt RS , Hatfield JL, Sauer TJ. Wind erosion. Soil Management: Building a Stable Base for Agriculture. 2011;**1409**:209-227

[20] Nearing MA, Xie Y, Liu B, Ye Y. Natural and anthropogenic rates of soil erosion. International Soil and Water Conservation Research. 2017;**5**(2):77-84

[21] Grimm M, Jones R, Montanarella L. Soil Erosion Risk in Europe: Joint Research Centre. Netherlands: European Commission Sint Maartensvlotbrug; 2001

[22] Quillard T, Franck G, Mawson T, Folco E, Libby P. Mechanisms of erosion of atherosclerotic plaques. Current Opinion in Lipidology. 2017;**28**(5):434

[23] Thornes JB. Modelling soil erosion by grazing: Recent developments and new approaches. Geographical Research. 2007;**45**(1):13-26

[24] Lal R. Soil management and restoration for C sequestration to mitigate the accelerated greenhouse effect. Environmental Sciences Proceedings; 1999;**1**:307-326

[25] Seutloali KE, Beckedahl HR. A review of road-related soil erosion: An assessment of causes, evaluation techniques and available control measures. Earth Sciences Research Journal. 2015;**19**(1):73-80

[26] Wantzen KM, Mol JH. Soil erosion from agriculture and mining: A threat to tropical stream ecosystems. Agriculture. 2013;**3**(4):660-683

[27] García-Ruiz JM. The effects of land uses on soil erosion in Spain: A review. Catena. 2010;**81**(1):1-11

[28] Lal R, Moldenhauer WC. Effects of soil erosion on crop productivity. Critical Reviews in Plant Sciences. 1987;**5**(4):303-367

[29] Case N. Environmental Controls on Clogging in Effluent-Dominated. Waterways: Arizona State University; 2012

[30] Vallero DA. Fundamentals of Air Pollution. Academic press an imprint of Elsevier. The Boulevard, Lane, Kington, Oxford OX5 1GB, UK; 2014

[31] Squires VR, Glenn EP. Salination, Desertification and Soil Erosion. The role of food, Agriculture forestry and fisheries in human nutrition. 2011

[32] Brown LR, Wolf EC. Soil Erosion: Quiet Crisis in the World Economy. World watch Paper 60; Worldwide Institute, NW, Washington, DC. 20036. 1984

[33] Peters GM, Wiedemann S, Rowley HV, Tucker R, Feitz AJ, Schulz M. Assessing agricultural soil acidification and nutrient management in life cycle assessment. The International Journal of Life Cycle Assessment. 2011;**16**:431-441

[34] Zuazo VHO, Pleguezuelo CRR. Soil erosion and runoff prevention by plant covers: A review. Sustainable Agronomy for Sustainable Development. 2008;**28**:65-86

[35] Issaka S, Ashraf MA. Impact of soil erosion and degradation on water quality: A review. Geology, Ecology, and Landscapes. 2017;**1**(1):1-11

[36] Pimentel D, Burgess M. Soil erosion threatens food production. Agriculture. 2013;**3**(3):443-463

[37] Boardman J. Soil erosion and flooding on the eastern south downs, southern England, 1976-2001. Transactions of the Institute of British Geographers. 2003;**28**(2):176-196

[38] Rashmi I, Karthika K, Roy T, Shinoji K, Kumawat A, Kala S, et al. Soil erosion and sediments: A source of contamination and impact on agriculture productivity. Agrochemicals in Soil and Environment: Impacts and Remediation. Singapore: Springer; 2022;**152**:313-345

[39] Orgiazzi A, Panagos P. Soil biodiversity and soil erosion: It is time to get married: Adding an earthworm factor to soil erosion modelling. Global Ecology and Biogeography. 2018;**27**(10):1155-1167

[40] Johnson JM-F, Franzluebbers AJ, Weyers SL, Reicosky DC. Agricultural opportunities to mitigate greenhouse gas emissions. Environmental Pollution. 2007;**150**(1):107-124

[41] Lorenz M, Fürst C, Thiel E. A methodological approach for deriving regional crop rotations as basis for the assessment of the impact of agricultural strategies using soil erosion as example. Journal of Environmental Management. 2013;**127**:S37-S47

[42] Sadeghi SHR. Soil erosion in Iran: State of the art, tendency and solutions. Poljoprivreda i Sumarstvo. 2017;**63**(3):33-37

[43] Shanks LW, Moore DE, Sanders CE. Soil erosion. Cover Cropping in Vineyards A GROWER's Handbook. University of California Division of Agriculture Recourses Publication 3338. 1998;**1**:80-85

[44] Tsoraeva E, Alborova P, Bazaeva L, Khanaeva D, Kozyrev B. Modern innovative and unconventional methods to combat soil erosion. In: E3S Web of Conferences. EDP Sciences; Topical Problems of Green Architecture, Civil and Environmental Engineering. 2021

[45] Mitas L, Mitasova H. Distributed soil erosion simulation for effective erosion prevention. Water Resources Research. 1998;**34**(3):505-516

[46] Gómez JA, Infante-Amate J, González de Molina M, Vanwalleghem T, Taguas EV, Lorite I. Olive cultivation, its impact on soil erosion and its progression into yield impacts in southern Spain in the past as a key to a future of increasing climate uncertainty. Agriculture. 2014;**4**(2):170-198

[47] Owens PN. Soil erosion and sediment dynamics in the Anthropocene: A review of human impacts during a period of rapid global environmental change. Journal of Soils and Sediments. 2020;**20**:4115-4143

[48] Borrelli P, Alewell C, Alvarez P, Anache JAA, Baartman J, Ballabio C, et al. Soil erosion modelling: A global review and statistical analysis. Science of the Total Environment. 2021;**780**:146494

[49] Luvai A, Obiero J, Omuto C, Sang J. Soil erosion and sediment yield modeling for the Maruba dam catchment, Machakos County, Kenya. Modeling Earth Systems and Environment. 2022;**8**(4):5723-5742

[50] Rajani B, Kleiner Y. Comprehensive review of structural deterioration of water mains: Physically based models. Urban Water. 2001;**3**(3):151-164

[51] Prasannakumar V, Vijith H, Abinod S, Geetha N. Estimation of soil erosion risk within a small mountainous

sub-watershed in Kerala, India, using revised universal soil loss equation (RUSLE) and geo-information technology. Geoscience Frontiers. 2012;**3**(2):209-215

[52] Morgan RPC. Soil Erosion and Conservation. John Wiley & Sons; The United Kingdom by MPG Books Ltd, Bodmin, Cornwall. 2009

[53] Van LN, Le X-H, Nguyen GV, Yeon M, Jung S, Lee G. Investigating behavior of six methods for sediment transport capacity estimation of spatial-temporal soil erosion. Water. 2021;**13**(21):3054

[54] Baensch-Baltruschat B, Kocher B, Kochleus C, Stock F, Reifferscheid G. Tyre and road wear particles-a calculation of generation, transport and release to water and soil with special regard to German roads. Science of the Total Environment. 2021;**752**:141939

[55] Jarrah M, Mayel S, Tatarko J, Funk R, Kuka K. A review of wind erosion models: Data requirements, processes, and validity. Catena. 2020;**187**:104388

[56] Tesfahunegn GB. Soil Erosion Modeling and Soil Quality Evaluation for Catchment Management Strategies in Northern Ethiopia. Bonn: Universitäts- und Landesbibliothek Bonn; 2011

[57] Foster G. A closed-form soil erosion equation for upland areas. Sedimentation. 1972;**1**:12-19

[58] Hewawasam T. Effect of land use in the upper Mahaweli catchment area on erosion, landslides and siltation in hydropower reservoirs of Sri Lanka. Journal of the National Science Foundation of Sri Lanka. 2010;**38**(1):3-14

[59] Foster GR, Meyer L, Onstad C. An erosion equation derived from basic erosion principles. Transactions of the ASAE. 1977;**20**(4):678-0682

[60] Walling D. Recent Advances in the Use of Environmental Radionuclides in Soil Erosion Investigations. 2003:57-78

[61] Ashraf A, Abuzar MK, Ahmad B, Ahmad MM, Hussain Q. Modeling risk of soil erosion in high and medium rainfall zones of Pothwar region, Pakistan: Assessment of soil erosion risk. Proceedings of the Pakistan Academy of Sciences: B Life and Environmental Sciences. 2017;**54**(2):67-77

[62] Abuzar MK, Shakir U, Ashraf MA, Mukhtar R, Khan S, Shaista S, et al. GIS based risk modeling of soil erosion under different scenarios of land use change in Simly watershed of Pakistan. Journal of Himalayan Earth Sciences. 2018;**51**(2A):132-143

[63] Jiu J, Wu H, Li S. The implication of land-use/land-cover change for the declining soil erosion risk in the three gorges reservoir region, China. International Journal of Environmental Research and Public Health. 2019;**16**(10):1856

[64] Harmon RS, Doe WW III, Doe WW. Landscape Erosion and Evolution Modeling. New York, Boston, Dordrecht, London, Moscow: Springer Science & Business Media; 2001

[65] Ighodaro ID, Lategan FS, Yusuf SF. The impact of soil erosion on agricultural potential and performance of Sheshegu community farmers in the eastern cape of South Africa. Journal of Agricultural Science. 2013;**5**(5):140

[66] Koirala P, Thakuri S, Joshi S, Chauhan R. Estimation of soil erosion in Nepal using a RUSLE modeling and geospatial tool. Geosciences. 2019;**9**(4):147

[67] Renschler CS, Harbor J. Soil erosion assessment tools from point to regional

scales—The role of geomorphologists in land management research and implementation. Geomorphology. 2002;**47**(2-4):189-209

[68] Finlayson DP, Montgomery DR. Modeling large-scale fluvial erosion in geographic information systems. Geomorphology. 2003;**53**(1-2):147-164

[69] Yesuph AY, Dagnew AB. Land use/cover spatiotemporal dynamics, driving forces and implications at the Beshillo catchment of the Blue Nile Basin, north eastern highlands of Ethiopia. Environmental Systems Research. 2019;**8**(1):1-30

[70] Habib-ur-Rehman S, Khan H, Khan S, Ahmad N, Bhatti WM. Incidence and gross pathology of salmonellosis in chicken in Hyderabad. Journal of Animal and Veterinary Advances. 2003;**2**:581-584

[71] Ismail J, Ravichandran S. RUSLE2 model application for soil erosion assessment using remote sensing and GIS. Water Resources Management. 2008;**22**:83-102

[72] Rewerts CC, Engel B. ANSWERS on GRASS: Integrating a Watershed Simulation with a GIS. Paper-American Society of Agricultural Engineers (USA) no, Chicago, Illinois. Availability. 1991. 91-2621

[73] Mitchell J, Engel B, Srinivasan R, Wang S. Validation of AGNPS for small watersheds using an integrated AGNPS/GIS system 1. JAWRA Journal of the American Water Resources Association. 1993;**29**(5):833-842

[74] Gaffer RL, Flanagan D, Denight ML, Engel BA. Geographical information system erosion assessment at a military training site. Journal of Soil and Water Conservation. 2008;**63**(1):1-10

[75] Wischmeier WH, Smith DD. Predicting Rainfall Erosion Losses: A Guide to Conservation Planning. Department of Agriculture Science and Education Administration; The Secretary of Agriculture, Washington, D.C. 20259. 1978

[76] Moore ID, Wilson JP. Length-slope factors for the revised universal soil loss equation: Simplified method of estimation. Journal of Soil and Water Conservation. 1992;**47**(5):423-428

[77] Mitasova H, Hofierka J, Zlocha M, Iverson LR. Modelling topographic potential for erosion and deposition using GIS. International Journal of Geographical Information Systems. 1996;**10**(5):629-641

[78] Desmet P, Govers G. A GIS procedure for automatically calculating the USLE LS factor on topographically complex landscape units. Journal of Soil and Water Conservation. 1996;**51**(5):427-433

[79] Cebecauer T, Hofierka J. The consequences of land-cover changes on soil erosion distribution in Slovakia. Geomorphology. 2008;**98**(3-4):187-198

[80] Pandey A, Chowdary V, Mal B. Sediment yield modelling of an agricultural watershed using MUSLE, remote sensing and GIS. Paddy and Water Environment. 2009;**7**:105-113

[81] Pandey A, Mathur A, Mishra SK, Mal B. Soil erosion modeling of a Himalayan watershed using RS and GIS. Environmental Earth Sciences. 2009;**59**:399-410

[82] Jain MK, Das D. Estimation of sediment yield and areas of soil erosion and deposition for watershed prioritization using GIS and remote sensing. Water Resources Management. 2010;**24**(10):2091-2112

[83] Lu D, Li G, Valladares GS, Batistella M. Mapping soil erosion risk in Rondonia, Brazilian Amazonia: Using RUSLE, remote sensing and GIS. Land Degradation & Development. 2004;**15**(5):499-512

[84] Dziewonski A, Hales A, Lapwood E. Parametrically simple earth models consistent with geophysical data. Physics of the Earth and Planetary Interiors. 1975;**10**(1):12-48

[85] Cox C, Madramootoo C. Application of geographic information systems in watershed management planning in St. Lucia. Computers and Electronics in Agriculture. 1998;**20**(3):229-250

[86] Molnár DK, Julien PY. Estimation of upland erosion using GIS. Computers & Geosciences. 1998;**24**(2):183-192

[87] Neshat A, Pradhan B, Pirasteh S, Shafri HZM. Estimating groundwater vulnerability to pollution using a modified DRASTIC model in the Kerman agricultural area, Iran. Environmental Earth Sciences. 2014;**71**:3119-3131

[88] Erdogan EH, Erpul G, Bayramin İ. Use of USLE/GIS methodology for predicting soil loss in a semiarid agricultural watershed. Environmental Monitoring and Assessment. 2007;**131**:153-161

[89] Wang G, Gertner G, Fang S, Anderson AB. Mapping multiple variables for predicting soil loss by geostatistical methods with TM images and a slope map. Photogrammetric Engineering & Remote Sensing. 2003;**69**(8):889-898

[90] Li X, Wu B, Wang H, Zhang J. Regional soil erosion risk assessment in Hai Basin. Yaogan Xuebao- Journal of Remote Sensing. 2011;**15**(2):372-387

[91] Chalise D, Kumar L, Shriwastav CP, Lamichhane S. Spatial assessment of soil erosion in a hilly watershed of Western Nepal. Environmental Earth Sciences. 2018;**77**:1-11

[92] Renard KG, Foster GR, Weesies GA, Porter JP. RUSLE: Revised universal soil loss equation. Journal of Soil and Water Conservation. 1991;**46**(1):30-33

[93] Shamshad A, Azhari M, Ma I, Hussin WW, Parida B. Development of an appropriate procedure for estimation of RUSLE EI30 index and preparation of erosivity maps for Pulau Penang in peninsular Malaysia. Catena. 2008;**72**(3):423-432

[94] Woldemariam GW, Iguala AD, Tekalign S, Reddy RU. Spatial modeling of soil erosion risk and its implication for conservation planning: The case of the Gobele watershed, east Hararghe zone, Ethiopia. Land. 2018;7(1):25

[95] Igbokwe J, Akinyede J, Dang B, Alaga T, Ono M, Nnodu V, et al. Mapping and monitoring of the impact of gully erosion in southeastern Nigeria with satellite remote sensing and geographic information system. The International Archives of the Photogrammetry, Remote Sensing and Spatial Information Sciences. 2008;**37**(Part B8):865-872

[96] Okereke C, Onu N, Akaolisa C, Ikoro D, Ibeneme S, Ubechu B, et al. Mapping gully erosion using remote sensing technique: A case study of Okigwe area, southeastern Nigeria. International Journal of Engineering Research and Applications. 2012;**2**(3):1955-1967

[97] Shikangalah R, Paton E, Jetlsch F, Blaum N. Quantification of areal extent of soil erosion in dryland urban areas: An example from Windhoek, Namibia. Cities and the Environment (CATE). 2017;**10**(1):8

[98] Opeyemi OA, Abidemi FH, Victor OK. Assessing the impact of soil

erosion on residential areas of Efon-Alaaye Ekiti, Ekiti-state, Nigeria. Journal of Environmental Planning and Management. 2019;**5**(9):22-31

[99] Tshikeba Kabantu M, Muamba Tshimanga R, Onema Kileshye JM, Gumindoga W, Tshimpampa BJ. A GIS-based estimation of soil erosion parameters for soil loss potential and erosion hazard in the city of Kinshasa, the Democratic Republic of Congo. Proceedings of the International Association of Hydrological Sciences. 2018;**378**:51-57

[100] Khosrokhani M, Pradhan B. Spatio-temporal assessment of soil erosion at Kuala Lumpur metropolitan city using remote sensing data and GIS. Geomatics, Natural Hazards and Risk. 2014;**5**(3):252-270

[101] Renard K, Foster G, Weesies G, McCool D, Yoder D. Predicting soil erosion by water: A guide to conservation planning with the revised universal soil loss equation (RUSLE). In: Agriculture Handbook. 1996;**537**:1-384

[102] Zhou P, Luukkanen O, Tokola T, Nieminen J. Effect of vegetation cover on soil erosion in a mountainous watershed. Catena. 2008;**75**(3):319-325

[103] Cerri CE, Demattě JA, Ballester MV, Martinelli LA, Victoria RL, Roose E. GIS erosion risk assessment of the Piracicaba River basin, southeastern Brazil. Mapping Sciences and Remote Sensing. 2001;**38**(3):157-171

[104] Al-Abadi AMA, Ghalib HB, Al-Qurnawi WS. Estimation of soil erosion in northern Kirkuk governorate, Iraq using rusle, remote sensing and gis. Carpathian Journal of Earth and Environmental Sciences. 2016;**11**(1):153-166

[105] Tundu C, Tumbare MJ, Kileshye Onema J-M. Sedimentation and its impacts/effects on river system and reservoir water quality: Case study of Mazowe catchment, Zimbabwe. Proceedings of the International Association of Hydrological Sciences. 2018;**377**:57-66

[106] Karamage F, Zhang C, Liu T, Maganda A, Isabwe A. Soil erosion risk assessment in Uganda. Forests. 2017;**8**(2):52

[107] Behera M, Sena DR, Mandal U, Kashyap PS, Dash SS. Integrated GIS-based RUSLE approach for quantification of potential soil erosion under future climate change scenarios. Environmental Monitoring and Assessment. 2020;**192**:1-18

[108] Renard KG. Predicting Soil Erosion by Water: A Guide to Conservation Planning with the Revised Universal Soil Loss Equation (RUSLE). United States Government Printing; Unite State Department of Agriculture. 1997

[109] Jahun B, Ibrahim R, Dlamini N, Musa S. Review of soil erosion assessment using RUSLE model and GIS. Journal of Biology, Agriculture and Healthcare. 2015;**5**(9):36-47

[110] Han F, Ren L, Zhang X, Li Z. The WEPP model application in a small watershed in the loess plateau. PLoS One. 2016;**11**(3):e0148445

[111] Ganasri B, Ramesh H. Assessment of soil erosion by RUSLE model using remote sensing and GIS-A case study of Nethravathi Basin. Geoscience Frontiers. 2016;**7**(6):953-961

[112] Smith H. Application of empirical soil loss models in southern Africa: A review. South African Journal of Plant and Soil. 1999;**16**(3):158-163

[113] Lilly A, Grieve I, Jordan C, Baggaley N, Birnie R, Futter M, et al.,

Climate change, land management and erosion in the organic and organo-mineral soils in Scotland and Northern Ireland. 2009.

[114] Tamene L, Vlek PL. Soil erosion studies in northern Ethiopia. Land Use and Soil Resources. Dordrecht: Springer; 2008;**180**:73-100

[115] Sharpley AN, McDowell RW, Kleinman PJ. Phosphorus loss from land to water: Integrating agricultural and environmental management. Plant and Soil. 2001;**237**:287-307

[116] Layek J, Das A, Ghosh PK, Rangappa K, Lal R, Idapuganti RG, et al. Double no-till and rice straw retention in terraced sloping lands improves water content, soil health and productivity of lentil in Himalayan foothills. Soil and Tillage Research. 2022;**221**:105381

[117] Roper M, Davies S, Blackwell P, Hall D, Bakker D, Jongepier R, et al. Management options for water-repellent soils in Australian dryland agriculture. Soil Research. 2015;**53**(7):786-806

[118] Adcock D, McNeill A, McDonald G, Armstrong R. Subsoil constraints to crop production on neutral and alkaline soils in south-eastern Australia: A review of current knowledge and management strategies. Australian Journal of Experimental Agriculture. 2007;**47**(11):1245-1261

[119] Singer MJ, Shainberg I. Mineral soil surface crusts and wind and water erosion. Earth Surface Processes and Landforms: The Journal of the British Geomorphological Research Group. 2004;**29**(9):1065-1075

[120] Allen V, Baker M, Segarra E, Brown C. Integrated irrigated crop–livestock systems in dry climates. Agronomy Journal. 2007;**99**(2):346-360

[121] Sharpley AN, Weld JL, Beegle DB, Kleinman PJ, Gburek W, Moore P, et al. Development of phosphorus indices for nutrient management planning strategies in the United States. Journal of Soil and Water Conservation. 2003;**58**(3):137-152

[122] de Paul OV, Lal R. Assessing land cover and soil quality by remote sensing and geographical information systems (GIS). Catena. 2013;**104**:77-92

[123] Servenay A, Prat C. Erosion extension of indurated volcanic soils of Mexico by aerial photographs and remote sensing analysis. Geoderma. 2003;**117**(3-4):367-375

[124] Kheir RB, Cerdan O, Abdallah C. Regional soil erosion risk mapping in Lebanon. Geomorphology. 2006;**82**(3-4):347-359

[125] Fadul HM, Salih AA, Imad-eldin AA, Inanaga S. Use of remote sensing to map gully erosion along the Atbara River, Sudan. International Journal of Applied Earth Observation and Geoinformation. 1999;**1**(3-4):175-180

[126] Nusser SM, Goebel JJ. The National Resources Inventory: A long-term multi-resource monitoring programme. Environmental and Ecological Statistics. 1997;**4**(3):181

[127] China MoWRotPsRo. Standards for Classification and Gradation of Soil Erosion. People's Republic of China Water Industry Standard SL; 2008. pp. 190-2007

[128] Tian Y, Zhou Y, Wu B, Zhou W. Risk assessment of water soil erosion in upper basin of Miyun reservoir, Beijing, China. Environmental Geology. 2009;**57**:937-942

[129] Zhang X, Wu B, Ling F, Zeng Y, Yan N, Yuan C. Identification of priority

areas for controlling soil erosion. Catena. 2010;**83**(1):76-86

[130] Panagos P, Meusburger K, Van Liedekerke M, Alewell C, Hiederer R, Montanarella L. Assessing soil erosion in Europe based on data collected through a European network. Soil Science and Plant Nutrition. 2014;**60**(1):15-29

[131] Prasannakumar V, Shiny R, Geetha N, Vijith H. Spatial prediction of soil erosion risk by remote sensing, GIS and RUSLE approach: A case study of Siruvani river watershed in Attapady valley, Kerala, India. Environmental Earth Sciences. 2011;**64**:965-972

[132] Da Silva RM, Montenegro SMGL, Santos CAG. Integration of GIS and remote sensing for estimation of soil loss and prioritization of critical sub-catchments: A case study of Tapacurá catchment. Natural Hazards. 2012;**62**:953-970

[133] Tsara M, Kosmas C, Kirkby M, Kosma D, Yassoglou N. An evaluation of the PESERA soil erosion model and its application to a case study in Zakynthos, Greece. Soil Use and Management. 2005;**21**(4):377-385

[134] Laflen JM, Moldenhauer W. Pioneering Soil Erosion Prediction: The USLE Story. World Association of Soil and Water Conservation. Beijing; 2003

[135] Wischmeier WH. Use and misuse of the universal soil loss equation. Journal of Soil and Water Conservation. 1976;**31**(1):5-9

[136] Kirkby M, Irvine B, Jones RJ, Govers G, Team P. The PESERA coarse scale erosion model for Europe. I.– model rationale and implementation. European Journal of Soil Science. 2008;**59**(6):1293-1306

[137] Tiwari A, Risse L, Nearing M. Evaluation of WEPP and its comparison with USLE and RUSLE. Transactions of the ASAE. 2000;**43**(5):1129-1135

[138] Stroosnijder L. Measurement of erosion: Is it possible? Catena. 2005;**64**(2-3):162-173

[139] Lal R. Soil degradation by erosion. Land Degradation & Development. 2001;**12**(6):519-539

[140] Montanarella L, Pennock DJ, McKenzie N, Badraoui M, Chude V, Baptista I, et al. World's soils are under threat. Soil. 2016;**2**(1):79-82

[141] Borrelli P, Robinson DA, Fleischer LR, Lugato E, Ballabio C, Alewell C, et al. An assessment of the global impact of 21st century land use change on soil erosion. Nature Communications. 2017;**8**(1):2013

[142] Borrelli P, Robinson DA, Panagos P, Lugato E, Yang JE, Alewell C, et al. Land use and climate change impacts on global soil erosion by water (2015-2070). Proceedings of the National Academy of Sciences. 2020;**117**(36):21994-22001

[143] Pimentel D, Harvey C, Resosudarmo P, Sinclair K, Kurz D, McNair M, et al. Environmental and economic costs of soil erosion and conservation benefits. Science. 1995;**267**(5201):1117-1123

[144] Zheng F-L. Effects of accelerated soil erosion on soil nutrient loss after deforestation on the loess plateau. Pedosphere. 2005;**15**(6):707-715

[145] Bai Z, Dent DL, Olsson L, Schaepman ME. Global Assessment of Land Degradation and Improvement: 1. Identification by Remote Sensing. ISRIC-World Soil Information; 2008

[146] Zika M, Erb K-H. The global loss of net primary production resulting

from human-induced soil degradation in drylands. Ecological Economics. 2009;**69**(2):310-318

[147] Fenta AA, Tsunekawa A, Haregeweyn N, Poesen J, Tsubo M, Borrelli P, et al. Land susceptibility to water and wind erosion risks in the East Africa region. Science of the Total Environment. 2020;**703**:135016

[148] Li Y, Tan M, Hao H. The impact of global cropland changes on terrestrial ecosystem services value, 1992-2015. Journal of Geographical Sciences. 2019;**29**:323-333

[149] Mertz O, Ravnborg HM, Lövei GL, Nielsen I, Konijnendijk CC. Ecosystem services and biodiversity in developing countries. Biodiversity and Conservation. 2007;**16**:2729-2737

[150] Bouma J. Land quality indicators of sustainable land management across scales. Agriculture, Ecosystems & Environment. 2002;**88**(2):129-136

[151] Evans R. Assessment and monitoring of accelerated water erosion of cultivated land–when will reality be acknowledged? Soil Use and Management. 2013;**29**(1):105-118

[152] Nearing MA. Soil erosion and conservation. In: Environmental Modelling: Finding Simplicity in Complexity. 2013;**1**:365-378

[153] Nearing MA, Pruski FF, O'Neal MR. Expected climate change impacts on soil erosion rates: A review. Journal of Soil and Water Conservation. 2004;**59**(1):43-50

Chapter 2

Spatial Quantification of Soil Erosion Using Rusle Approach: A Study of Eastern Hindu Kush, Pakistan

Zara Tariq and Shakeel Mahmood

Abstract

Globally, soil erosion is a severe environmental issue, particularly in mountainous regions, leading to substantial declines in soil productivity. This study aims to quantify soil loss in Eastern Hindu Kush region using Revised Universal Soil Erosion Loss Equation (RUSLE) approach integrated with Remote Sensing and Geographic Information System (GIS). The study considers various factors including rainfall, soil erodibility, topography, slope, and land use to model annual soil loss rates. Rainfall erosivity (R), slope length and steepness (LS), soil erodibility (K), cover management (C), and conservation practice (P) were utilized as input parameters. These parameters are integrated to estimate soil erosion risk zones through raster-based GIS analysis, categorizing soil loss severity into five classes. The results show soil loss rates ranging from > 50 to over 276 tons/ha/year, indicating varying levels of severity. The distribution of soil loss severity is as follows: 37% of the area falls under insignificant, 16% under slight, 22% under moderate, 11% under severe, 6% under very severe, and 8% under catastrophic severity zones. Notably, valley areas with steep slopes and significant relief display higher erosion rates. The intricate and challenging terrain of the Eastern Hindu Kush makes it particularly susceptible to soil erosion risks.

Keywords: soil erosion, RUSLE, GIS, elevation, slope, susceptibility, Hindu Kush

1. Introduction

Soil degradation is one of the most leading environmental challenges all around the world. It decreases land for agricultural activities and ultimately lead to less agricultural production. It also leads in the removal of top soil which reduces the fertility rate of that land [1–3]. Soil is a medium which has been threatened by several factors like soil erosion, decrease in organic matter, contamination. The regions having arid and semi-arid climatic conditions are more prone towards soil erosion [4]. It has influence on land degradation, water quality, sedimentation of rivers, infrastructural damages and on agricultural productivity [5, 6]. The causes for erosion include, agricultural activities, urbanization, population explosion, climate change, infrastructure, mining activities and many other [7].

Short-term environmental and climatic change is resisted by most soils, but may undergo irreversible change such as large-scale erosion. Erosion of the soil is normally used to depict the adverse effects of man's utilization of soil resource, with soil being a valuable natural resource which is renewed really slowly. Soil erosion is induced by anthropogenic impact on land surface, whether in terms of deforestation, extreme cultivation or the misuse of the land. Rill and sheet erosion proves to be very dangerous form of soil erosion, resulting in an almost imperceptible but constant degradation of land under cultivation. Soil erosion proves to be disastrous natural phenomenon that threatens soil stability [8].

The identification and estimation of erosion risk zones is an important element in preventing land degradation. Among those elements the most important factors are the observation of soil forming factors of that particular region. In Pakistan, the phenomenon of soil degradation is also of utmost importance and is one of the major environmental challenges. Soil erosion has many impacts such as low agricultural productivity and sedimentation. The quantities of soil erosion depend upon the topography, vegetation, soil type, and climatic conditions. Pakistan is a dry land and lies in arid and semi-arid region. Eighty percent of the land in Pakistan is arid or semi-arid, about 12% is dry sub-humid and the remaining 8% is humid. About two-third population in Pakistan are depended on this dry land for their livelihood. Dry lands of Pakistan are drastically affected by degradation of land and desertification because of poor and mismanaged land practices [9].

Soil erosion assessment, zonation and prediction are highly important in order to lessen soil loss [10]. Several soil erosion modeling approaches have been introduced to predict soil erosion in highland regions and to evaluate the transportation and deposition of sediments. Among these models, majority of the models were first introduced in United States based entirely on different equations. Later on, these equations were improvised and many new variables and factors were added. Some of the soil erosion models are, Modified Universal Soil Loss Equation (MUSLE) [11], Universal Soil Loss Equation (USLE), the Unit Stream Power – based Erosion Deposition (USPED) [9, 12], and the Revised Universal Soil Loss Equation (RUSLE) [13].

The RUSLE is a computer-based version of USLE. It has been modified by adding several new factors which include new set of rules and algorithms to compute the cover factor, for slope length and steepness factors. This study aims to spatially quantify the soil loss in Eastern Hindu Kush Region using RUSLE approach integrated with Remote Sensing and Geographic Information System (GIS).

2. The study region

Geographically, Eastern Hindu Kush in Pakistan extends between 34° 34′11′′ to 36° 54′30′′ North Latitude and 71°11′56′′ to 73°52′5′′ East Longitude. Hindu Kush region lies in the west of Himalayas. Its western section falls in Afghanistan, however the eastern section located in Pakistan. The drainage basins of river Swat and Chitral are covering the eastern section of Hindu Kush mountain system. Swat River originates from two major glaciers of Ushu and Gabral, whereas river Chitral originates from Chiantar glacier. The study region is famous for its beautiful and fertile river valleys, which support large population (**Figure 1**) [14].

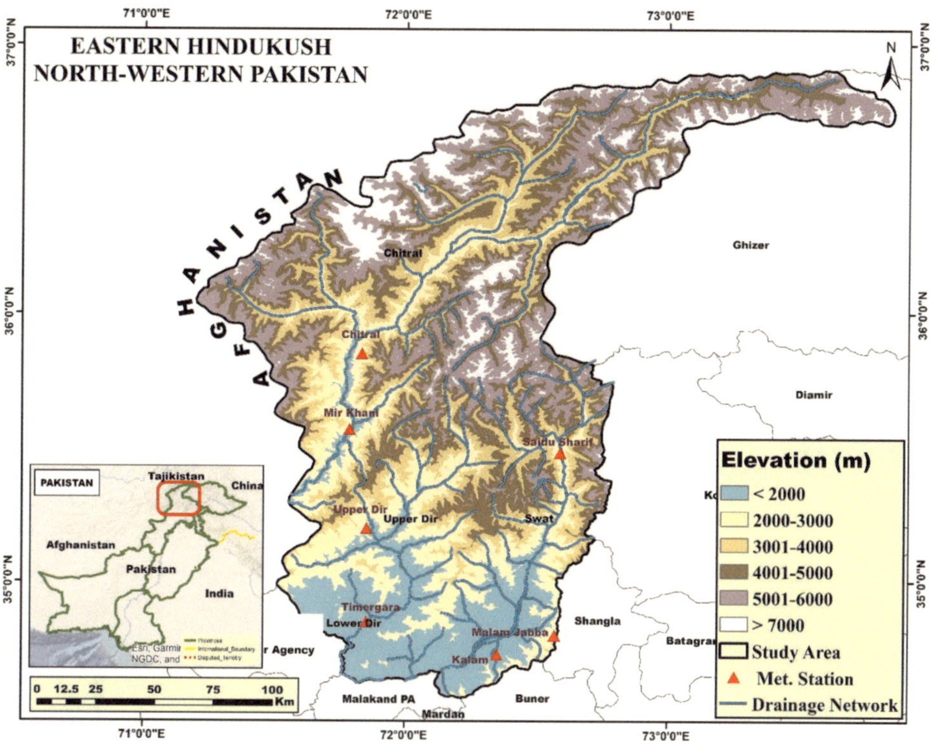

Figure 1.
The study area.

Eastern Hindu Kush stretches from Karambar Pass in East to the Dorah Pas not far from Mount Tirch Mir. The Central part continues to the Shebar Pass to the northwest of Kabul. The Western part of Hindu Kush descends to the Kermu Pass. In the extreme eastern region between the Karambar and Baroghil is dominated by very high peaks. Administratively, the Eastern Hindu Kush consists of the following districts, District of Chitral, Upper Dir, Lower Dir and Swat Scott [15]. The region is prone to heavy rainfall, surface runoff and flash floods which further intensity soil loss [16, 17].

3. Spatial quantification of soil loss

The extent of soil erosion estimation is a complex interactions between Geology, topography, climate, soil, land use and land cover. RUSLE approach was selected to predict protracted average annual soil loss rates in an area (**Figures 2** and **3**). In this model, five major parameters were utilized to quantify soil erosion loss in Eastern Hindu Kush region. The mathematical expression of RUSLE model is given in the following equation:

$$A = R \times K \times L \times S \times C \times P \quad (1)$$

Where,
A = Soil loss per unit area (tons/ha/yr.).
R = Rainfall-runoff erosivity factor (index) (MJ/hectare mm/yr.).

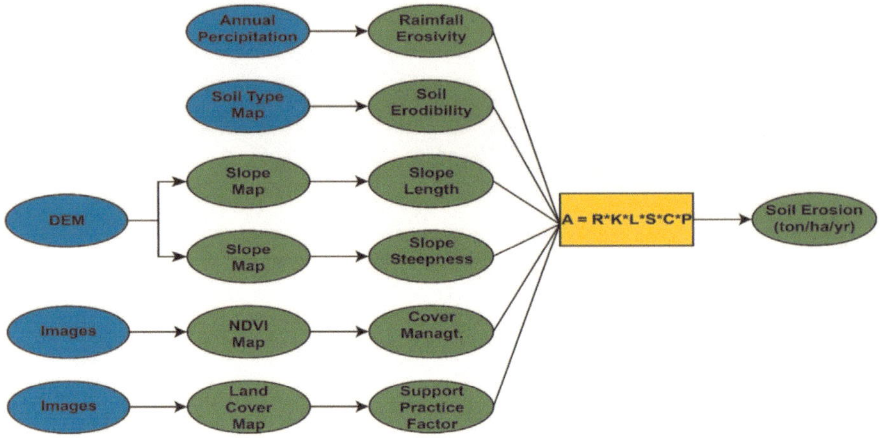

Figure 2.
RUSLE approach.

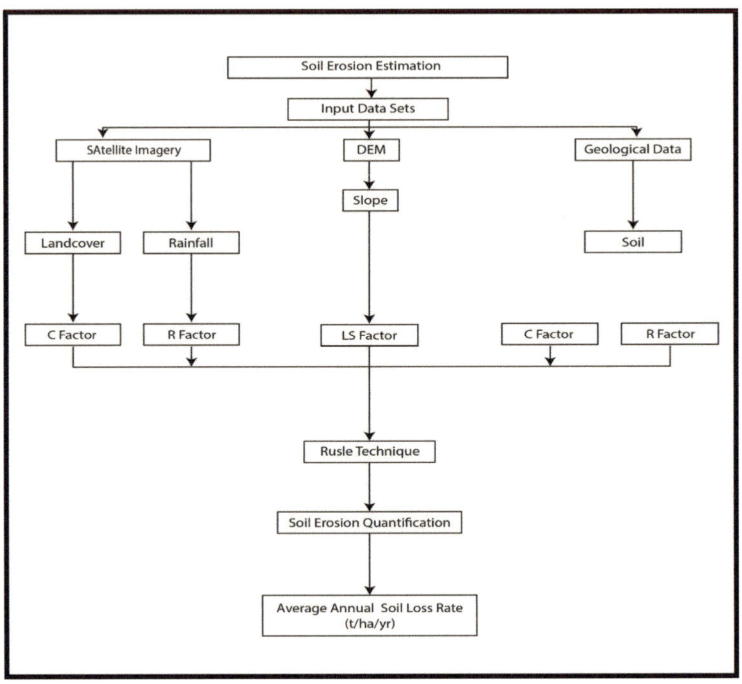

Figure 3.
Research work flow.

K = Soil erodibility factor (tons/ha/yr).
LS = Slope factor (unit less).
C = Cover management factor (unit less).
P = Conservation practice factor (unit less).

3.1 Rainfall-runoff Erosivity factor (R)

Rainfall-Runoff erosivity (R) quantifies the impact of raindrop on the surface and the rate of runoff likely to take place after a rain event. This factor is well-defined as the mean annual sum of individual or specific storm event energy (E), and also the maximum 30 min rainfall intensity for a specific storm event as described by [13, 18]. In order to estimate the accurate R factor, it is recommended to observe at least 20 to 30 years of rainfall data to accommodate climatic variation. The R factor determines the erosivity by rainfall at a specific region based on the intensity and amount of rainfall. It basically represents the impact of rainfall intensity on soil erosion. The rainfall-runoff erosivity was estimated by the following equation used by many researchers on areas where similar topographic and atmospheric conditions prevail [19, 20].

$$R = 0.05 \times P. \tag{2}$$

Where,
R = Rainfall Erosivity Factor, P = Mean Annual Rainfall in (mm).

3.2 Soil Erodibility factor (K)

The K factor is a quantitative measurement of the erodibility of a particular type of soil. It can be also described as a measure of the susceptibility of soil particles towards detachment and transportation by rainfall intensity and runoff. Soil texture, soil structure, soil permeability and the organic matter are the main soil properties influencing K factor. The soil erodibility factor for every particular soil is defined as the rate of erosion per unit erosion index from a standard unit plot of 22.13 m long slope length having 9% of slope gradient [21]. It represents the rate of soil loss per rainfall erosivity index (R).

On the basis of data availability, following equation was used to estimate the soil erodibility of soil given by Wischmeier and Smith [18].

$$K = \text{Fcsand}^* \ \text{Fsi} - \text{cl}^* \, \text{Forgc}^* \, \text{Fhisand}^* \, 0.1317. \tag{3}$$

Where,

$$\text{Fcsand} = \left[0.2 + 0.3 \ \exp\left(-0.0256 \ \text{SAN} \left(1 - \frac{SIL}{100}\right)\right)\right]. \tag{4}$$

$$Fsi - cl = \left(\frac{\text{SIL}}{\text{CLA} + \text{SIL}}\right) 0.3. \tag{5}$$

$$\text{Forgc} = \left(1.0 - \frac{0.25C}{C + \exp\{3.72 - 2.95C\}}\right). \tag{6}$$

$$\text{Fhisand} = \left(1.0 - \frac{0.70SNI}{SNI + \exp\{5.51 + 22.95SNI\}}\right). \tag{7}$$

Where, C is the organic carbon content, SIL, CLA and SAN are % silt, clay and sand, respectively, SN1 is sand content which is obtained by subtracting it from 1 and dividing by 100, Fcsand = gives a low soil erodibility factor for soil with coarse sand

and a high value for soil with little sand content, Fsi-cl gives a low soil erodibility factor with high clay to silt ration, Forgc is the factor that reduces soil erodibility for soil with high organic content, Fhisand is the factor that reduces soil erodibility for soil with extremely high sand content.

3.3 Slope length and slope steepness factor (LS)

The Slope length or Steepness factor (LS) is the output of two individual factors combined together i.e. Slope length factor (L) and a Slope gradient factor (S), both of these factors are delineated from the ALOS PALSAR DEM. The LS factor proves to be an important parameter in the modeling of soil erosion.

The L factor depicts impact of slope on soil erosion. When the length of slope increases, erosion of soil will also increase. Whereas, the S factor represents impact of slope gradient on erosion. The rate of soil loss increases with increasing slope steepness more than it does with length of slope. The LS factor depicts erodibility because of slope steepness and length. It signifies the influence of topography, specifically slope features, on soil erosion. Hence, proving it to be directly proportional to the soil erosion e.g., an increase in slope steepness and length marks an increase in the LS factor.

The LS factor was calculated from after generating the flow direction and flow accumulation grids in ArcMap 10.5 by using Arc Hydro toolset.

$$L = \left(\frac{\lambda}{22.13}\right)^{\mathrm{m}}. \qquad (8)$$

Where, L = Slope length factor, λ = Slope length (m), m = Slope-length exponent

$$m = \frac{F}{1+F'}. \qquad (9)$$

$$\frac{sin\,\beta/0.0896}{3\,(sin\,\beta)0.8 + 0.56}. \qquad (10)$$

Where, F = Ratio of rill erosion to inter-rill erosion, β = Slope angle (°).
In ArcMap, L was calculated by the following equation,

$$\mathrm{L} = \frac{(flow_{acc} + 625)^{(m+1)} - flow_{acc}{}^{(m+1)}}{25^{(m+2)} * 22.13^{m}} \qquad (11)$$

.

$$S = Con((Tan(\mathrm{slope} * 0.01745) < 0.09), (10.8 * \mathrm{Sin}(\mathrm{slop} * 0.01745) + 0.03), (16.8 * \mathrm{Sin}(\mathrm{slop} * 0.01745) - 0.5))$$

For Slope gradient factor,

$$S = Con((Tan(\mathrm{slope} * 0.01745) < 0.09), (10.8 * \mathrm{Sin}(\mathrm{slop} * 0.01745) + 0.03), (16.8 * \mathrm{Sin}(\mathrm{slop} * 0.01745) - 0.5)) \qquad (12)$$

.

Final LS Factor,

$$\mathrm{LS} = \mathrm{L}^{*}\,\mathrm{S}. \qquad (13)$$

3.4 Cover management factor (C)

Cover management factor (C) is used for estimation cropping impact and other managing practices on soil erosion. After topography, vegetation is considered the 2nd most vital aspect that helps in minimizing the risk of soil erosion. Different types of land use and land cover intercepts precipitation and increasing infiltration rates and also helps in the reduction of rainfall impact on ground by reducing its energy before hitting the ground.

In the study area, Global land cover data was used to generate a C-factor map. It was generated by modifying the dataset in a raster-based GIS environment. The shape file was then modified in ArcGIS by merging all the attributes of same grid codes of land cover type. The C values were assigned by reviewing the literature of comparable model usage in the areas having similar prevailing climatic conditions as my study area.

3.5 Erosion control practice management factor (P)

Erosion support practice factor (P) indicates the rate of soil loss according to different land cover management practices. This factor accounts for the control practices which reduces the rate of erosion caused by runoff and their influence on runoff concentration, runoff velocity, drainage patterns. P factor also accounts for the hydraulic forces exerted on soil by runoff. Land treatment in the form of contouring, strip cropping and terracing are the precautionary measures taken to prevent erosion. The precautionary measures or any control practices that are being used to minimize the impact of various factors on erosion contributes in the calculation of P factor.

The extent of soil erosion can be predicted by estimating the complex interactions between Geology, topography, climate, soil, land use and land cover. This empirical based technique is used globally to predict protracted average annual soil loss rates in an area. In this model, five major parameters are calculated to measure the soil erosion rates in a specific region (**Figure 3**). The work flow the Study is given below.

4. Spatial estimation of soil erosion

4.1 The R factor

R factor determines erosivity by rainfall at a specific region, which is usually based on the intensity and amount of precipitation. It mainly represents the impact caused by rainfall intensity on soil erosion. In this study area, the monthly precipitation data for six weather stations situated in the study area were obtained from the World Bank website. It was validated with the satellite based data to integrate the results. Rainfall map was prepared which represents the spatial distribution of rain in the Eastern Hindukush. This map was utilized to estimate R factor map by calculating the Rainfall erosivity for a time period of about 1991 to 2019 in study area. Rainfall-runoff erosivity was calculated by the following equation used by many researchers on areas where similar topographic and atmospheric conditions prevail (**Figure 4**) [19, 22].

$$R = 0.05 \times P. \qquad (14)$$

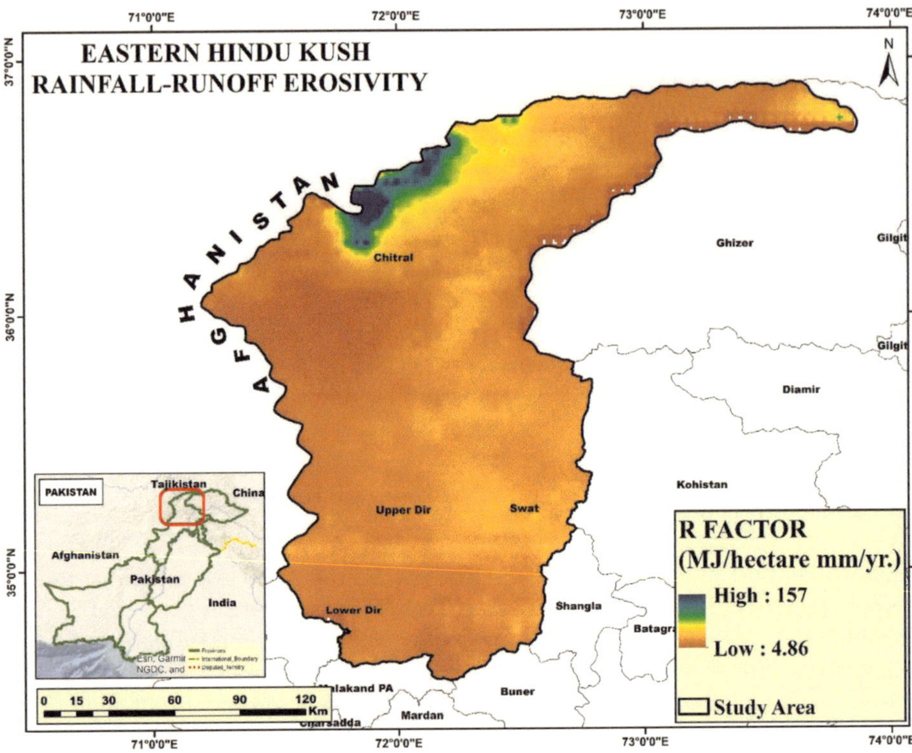

Figure 4.
Rainfall-Erosivity factor (R) of eastern Hindukush, Pakistan.

Where,
R = Rainfall Erosivity Factor, P = Mean Annual Rainfall in (mm).

4.2 Soil Erodibility factor (K)

K factor is the quantitative depiction of soil erodibility analyzed for a particular soil type. It basically is measure of soil particles susceptibility towards the impact of rainfall and runoff. The erodibility factor is mainly influenced by the texture of soil, soil structure, soil permeability and organic matter. For an individual soil type, K factor is particularly defined as the rate of soil erosion per unit erosion index measured by average unit plot of 22.13 m long. It chiefly determines the rate at which soil loss takes place per rainfall erosivity index (**Figures 5** and **6**).

4.3 Slope length and slope steepness factor (LS)

Slope length or Steepness factor, The LS factor is the output of 2 individual factors combined together i.e. Slope length factor (L) and a Slope steepness factor (S), both of these factors were delineated using the ALOS PALSAR DEM. The LS factor proves to be an important parameter in the modeling of soil erosion risk in eastern Hindukush region.

L factor depicts the abrupt impact of slope on soil erosion. The soil loss increases per unit area as the slope length increase. Whereas, the S factor represents the impact

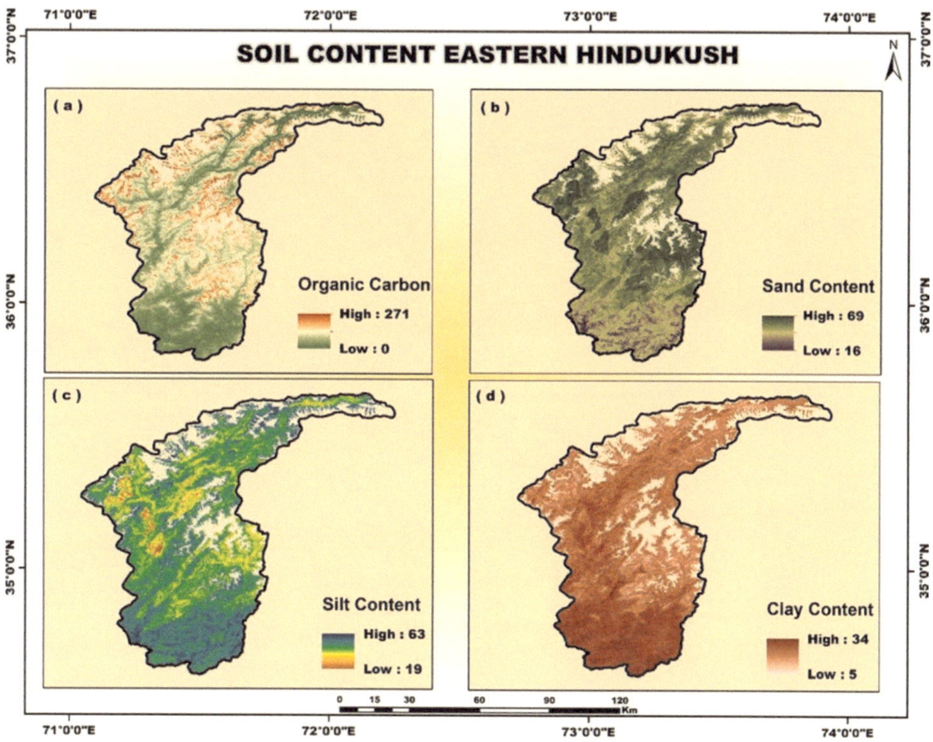

Figure 5.
Soil content.

of slope steepness on erosion. The rate of soil loss intensifies with increasing steepness of slope more than it does with length of slope. LS factor depicts erodibility because of slope steepness and length. It signifies the influence of topography, specifically slope features, on soil erosion. Hence, proving it to be directly proportional to the soil erosion for example, an increase in slope steepness and length marks an immense increase in the LS factor.

Slope length and steepness for the study area was calculated by utilizing an elevation model. The DEM was filled to fill all the depressions to get accuracy in the imminent analysis. When all the depressions or sinks formed because of erroneous data are filled, by assigning them the values of neighboring cells (**Figure 7**) [23].

4.4 Cover management factor (C)

The Cover management usually known as the C factor is used to estimate the impact of several management and cropping practices on soil erosion. After topography, vegetation is considered as the 2ndmostsignificant factor that helps in minimizing the risk of soil erosion. Different types of LULC intercepts precipitation and results in increasing infiltration rates and also helps in the reduction of rainfall impact on ground by reducing its energy before hitting the ground (**Figure 8**).

In the study area, Global land cover data was used to calculate and analyze a C-factor map. It was generated by converting the dataset into a polygon .shp. The shape file was then modified in ArcGIS by merging all the attributes of same grid codes of land cover type. Now, the C values were assigned by reviewing the literature

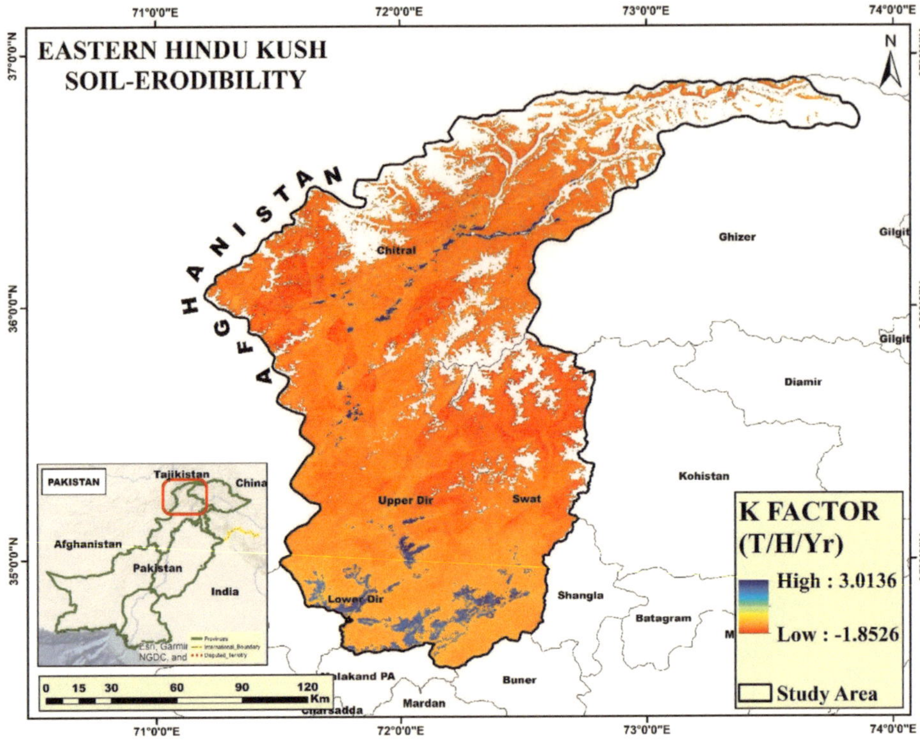

Figure 6.
Soil Erodibility factor (K) of eastern Hindukush, Pakistan.

of comparable model usage in the areas having similar prevailing climatic conditions (**Table 1**).

4.5 Erosion control practice management factor (P)

The P factor or erosion support practice factor indicates the rate of soil loss as per different land cover management practices. This factor particularly accounts for the control practices which reduces the rate of erosion caused by runoff, and also focusses their influence on drainage patterns, runoff concentration and runoff velocity. P factor also accounts for the hydraulic forces exerted on soil by runoff. Land treatment in the form of contouring, strip cropping and terracing are the precautionary measures taken to prevent erosion (**Figure 9**).

The precautionary measures or any control practices that are being used to minimize the impact of various elements on erosion contributes in the generation of P factor (**Table 2**).

This issue has never been addressed in the study region. Accordingly, no resistance or management strategies and techniques have been used. Consequently, the value of '1' was generally set for the generalization of the P factor in the eastern Hindukush, Pakistan.

5. Annual soil loss

The final soil loss magnitude was obtained by analyzing five Geo-environmental factors, Rainfall-Runoff Erosivity factor (R), Soil-Erodibility factor (K), Slope length

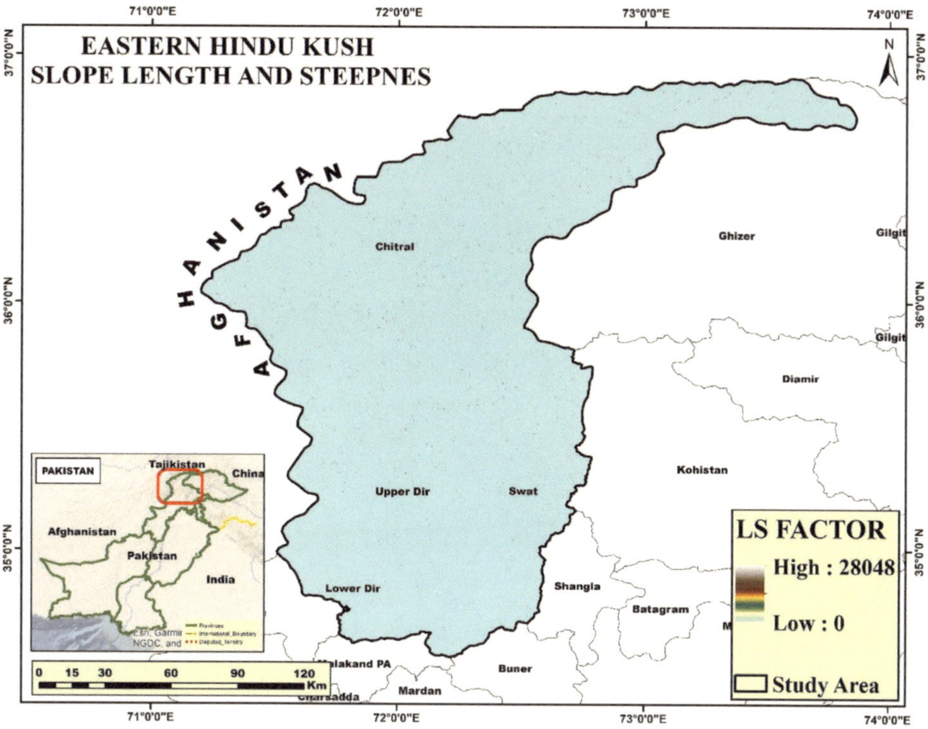

Figure 7.
Slope length and slope steepness (LS) of eastern Hindukush, Pakistan.

(L) and Slope Steepness (S), Cover management factor (C) and the Erosion Practice Control management factor (P).

The estimated annual average soil loss in the Eastern Hindukush is ranging from 50 to more than 276 Tons/ha/year. The spatial distribution of soil loss severity is represented in **Figure 10**. The study area is delineated into five different zones showing the severity of soil erosion. Maximum erosion is observed in northern parts of the study area. Bare areas and highlands with steep slopes are more susceptible to soil loss. Bare areas and highlands with steep slopes are more vulnerable to soil loss as shown in the map.

6. Discussion

Eastern Hindukush is mostly fed by glaciated water. When glaciers melt, it causes rapid erosion. This increased rate of erosion can affect the productivity of Dam which is located on Chitral River, leading to reduced power generation as well as loss of agriculture land. Initial output of the dam was about 250,00 kilowatts but with the passage of time, it reduced to 64,000–20,000 kilowatts.

RUSLE modeling approach comprises of following parameters; The Rainfall-Runoff Erosivity (R), Soil Erodibility (K), Slope Length and Slope Steepness (LS), Cover Management (C) and Support Practice (P) Factor. For the calculation of annual average soil erosion, the above five factors were estimated. Input layers of these factors were in ArcGIS individually. The high R factor values shows that the northern part of District Chitral receive more rainfall, as well as areas of Shoghar, Kalash and

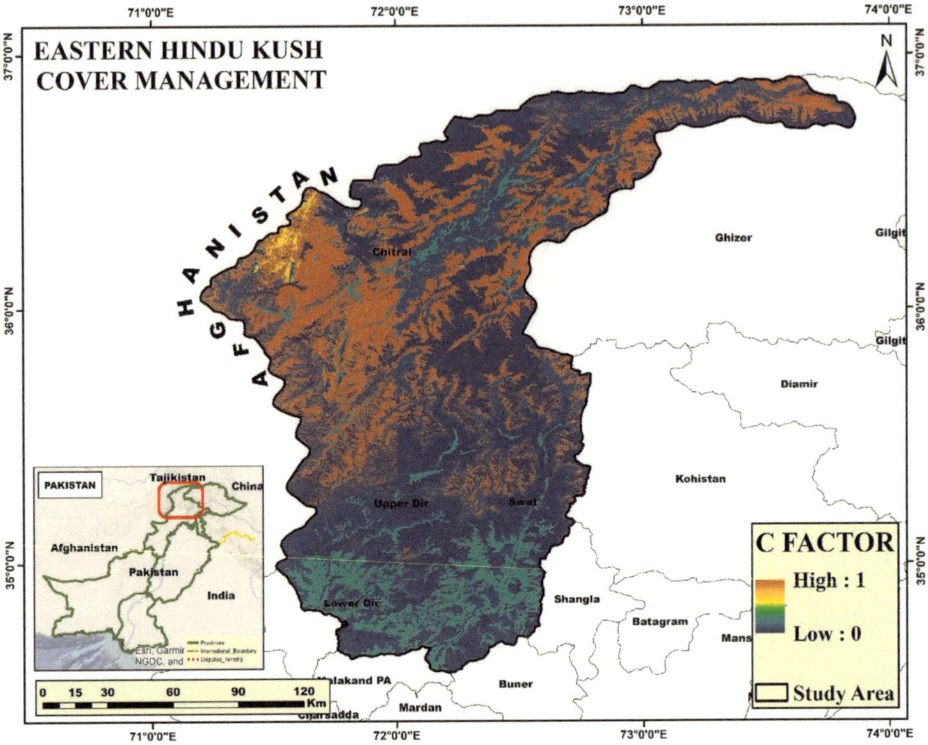

Figure 8.
Cover management (C) of eastern Hindukush, Pakistan.

Sr. no.	Land Cover	C Factor
1	Grass Land	0.059
2	Built-up Areas	0
3	Barren Land	1
4	Shrub Land	0.69
5	Water Bodies	0
6	Cultivated Land	0.28
7	Forest	0.004
8	Wetland	0
9	Snow Covered	0

Table 1.
C-factor in the eastern Hindukush.

the south eastern part of the study area as depicted by the results. The high elevation and topographical features of this region is the cause of excessive rain apart from other regions. The K factor of the study area ranges between −1 to 3.0. Soil erodibility map was generated using the ISRIC Soil Data of organic carbon, Clay, silt an = d sand content found in the region. Lower values of K factor show the soils having low

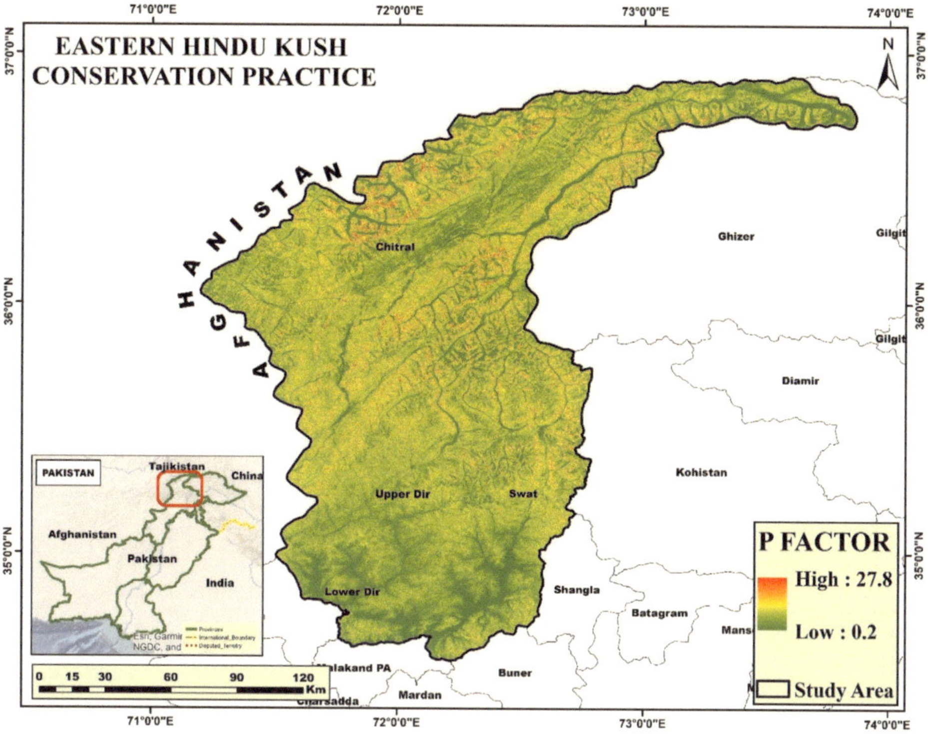

Figure 9.
Erosion control practice management (P) of eastern Hindukush, Pakistan.

Slope %	Contouring
0–7	0.55
7–11.3	0.6
11.3–17.6	0.8
17.6–26.8	0.95
> 26.8	1

Table 2.
P factor values for slope as per agricultural practice.

permeability and lower moisture content etc. The slope length and slope steepness (LS) factor has a range of 0 to 28,048. Flow accumulation and accelerated slope is represented by higher values of LS factor. Cover management, the C factor value falls in the range of 0 to 1.1 depicts the maximum land cover, whereas 0 represents the areas having minimum or very little land cover. The value of the support practice (P) factor is generated as 0.2 to 27.8. The region is mostly suffering from the ignorance of authorities on conservation, calculating this factor was a challenge because of the deficiency of data on conservation practices being carried out in the region. The annual average rate of soil loss of the study area is about 276 tons/ha/year, and the overall annual soil loss from this region is nearly 31 million tons/year. The results concluded by the severity classification of soil erosion estimated that % area of study

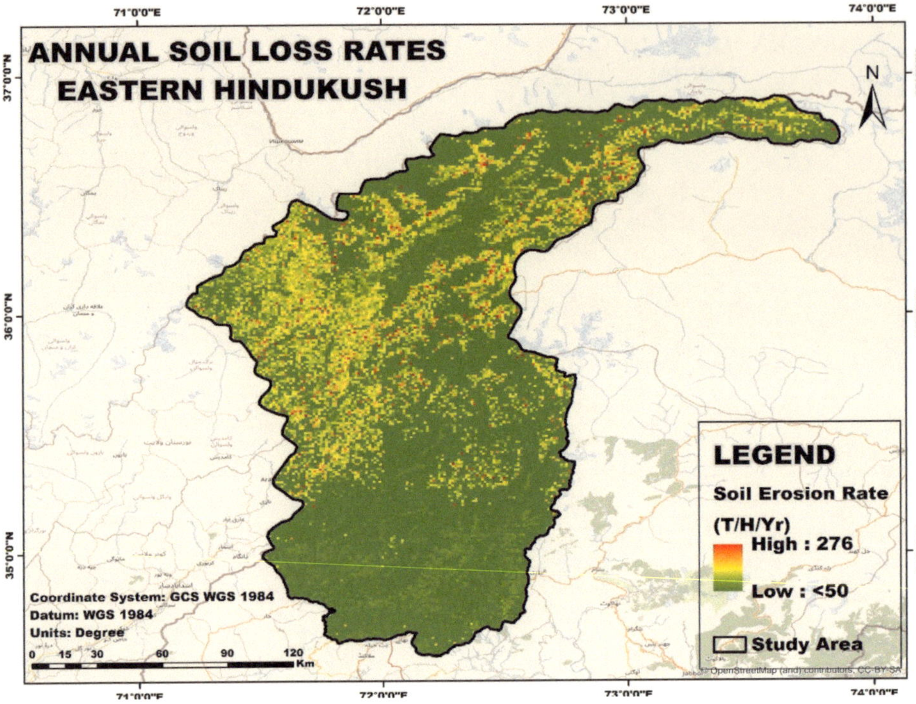

Figure 10.
Average annual soil loss rates (T/H/year) of the eastern Hindukush region, Pakistan.

area falls in the very low zone, indicating soil erosion of <10 tons/ha/year, which is contradictory to the bearable limit of <2 tons/ha/year [24]. It clearly shows that this region needs to be addressed properly and timely in this regard. Moreover, almost 30% of the area falls in the low-erosion zone, 22% of the area in the region lies in the moderate zone, 16% lies in high-erosion zones, whereas 8% of the total study area lies under catastrophic erosion zone. The areas having minimum vegetation and steep slopes are more vulnerable to high soil erosion. Several land use activities, such as urbanization, overgrazing and deforestation, considerably increase erosion rates, and make it critical in areas having steep slopes and high elevation.

RUSLE has proved to be efficient and suitable tool for soil loss estimation and is globally acknowledged because of its accurate and reliable calculation of annual soil loss rates [20]. On the other hand, practical validation was not achieved because the resources were insufficient. The final outcome of this study is comparable to formerly conducted studies in the neighboring regions at watershed level. The result of a study proposed in the watershed of Fateh Jang showed that soil loss was 17–41 tons/ha/year for 1–10% slopes in uncultivated land, while the rate was relatively lower (9–26 tons' ha/year) for vegetative land [20]. Another proposed research for the soil loss appraisal in plain areas using RUSLE showed almost 8 tons/ha/year. Results of the study conducted above showed that the Eastern Hindukush is at more risk to soil erosion than the Potohar region, Pakistan.

This study shows that soil erosion is a severe hazard and is in dire need to be addressed. This study will help in identification and understanding risks associated with soil erosion. It does not only identify the risk but also listed the contributory factors towards the erosion. The management and strategic planning to sustain natural resources and policymaking can analyze the findings of this study to minimizes oil loss.

7. Conclusion

This research provides the quantitative aspect of soil erosion. The research utilized RUSLE with GIS to map the soil erosion severity in the eastern Hindu Kush, Pakistan. The five parameters involved in the RUSLE model are rainfall–runoff erosivity (R), soil erodibility (K)) factor, slope length and steepness (LS) factor, land cover management (C) factor, and support practice (P) factor. These factors are used to estimate the annual average soil loss rate. The maps of these factors were generated separately, and value ranges were obtained from generated maps. The high value of rainfall–runoff erosivity factor (R) indicates that the northern part of the study area receives more rainfall, including the areas of Shoghar, Chitral, and Kalash, as indicated in the results. The topography and high elevation of this region is the major reason for the excessive rains. The soil erodibility (K) factor ranges between −1 to 3. The soil erodibility map was generated by using the raster soil content layers of the respective soil types in the soil map. The slope length (LS) factor has a range of 0 to 28,048. The higher value represents flow accumulation and an increase of slope. Land Cover management (C) factor lies in the range of 0 and 1. Maximum land cover is indicated by assigning it the value of 1, whereas the least land cover is represented by 0. The value of the support practice (P) factor is taken as 1.0 because of the ignorance of authorities on conservation and the deficiency of data on conservation practices.

The study concludes that it is quite significance to recognize and to have a complete understanding of risks related to the erosion of soil in the study area. The current study has proved Eastern Hindu Kush to be under severe threat of soil erosion. The highest rates of soil erosion are found along the path of river. River channel and some areas in Eastern Hindukush are highly vulnerable to soil erosion. Barren land and highlands with steep slopes are more vulnerable to soil loss. The lowermost portion of the Eastern Hindukush is prone to very high soil erosion rates. The magnitude of soil erosion was estimated by calculating the annual average soil loss rates in the Eastern Hindukush, which ranged between >50 to more than 276 Tons/ha/year presented by low and high values. Severity of the soil loss is represented by five different classes. Maximum erosion is observed in northern parts of the study area. Bare areas and highlands with steep slopes are more susceptible to Average annual soil loss. The percentages of the area lie under soil loss rates, concerning its severity, are 37% in insignificant, 16% in slight, 22% in moderate, 11% in severe 6% in very severe, and 8% in the catastrophic severity zone of the study area. Final output of the study was to calculate the soil erosion risk in the Eastern Hindu Kush region, Pakistan. The low and very low class represents areas having no or minimum risk towards soil erosion as compared to other classes. The areas bearing the moderate class are prone to soil erosion but the damages may not be catastrophic. High and very high classes represents the areas having maximum risk of erosion that can be catastrophic in nature if triggered in the near future.

This result of the study shows that the Eastern Hindukush is greatly prone to soil erosion, mainly the southern part of the region including the areas of Ispheru, Arkari, kalash and Harchin and many more. If the phenomenon of erosion expands with the same pace it would cause more land degradation. The results in this study comprise soil erosion severity classes and erosion intensity. The estimated soil loss in the present study was 276 tons/ha/year. The percentage of erosion which is about 8 percent shows that this region is very prone to erosion. The rate of soil erosion is increasing day by day; it will cause serious damage to the living conditions of eastern Hindukush. Thus, livelihoods of many families will be suffered if actions and precautionary measure are not taken in time. There are no precautionary measures in high elevated steep slopes to ensure the mitigation of soil erosion risk. Lack of knowledge and mismanaged agricultural practices in this region is also a major cause of soil erosion. This study will assist the policymakers and planners who can utilize these results to generate a mitigation strategy and can future planning as well.

8. Recommendations

Plantation in the upper parts of the area would be help in sustaining soil and it also will be helpful in stabilizing the climate. Moreover, plantation along the river channels will be helpful in protecting the agricultural land from flash floods resulting in erosion control. The construction of embankments along the river and adjacent to the river will prevent the land from erosion caused by flooding. Hydro-meteorological stations play an important role in collecting hydrological and meteorological observations, the modeling of hydrological and meteorological phenomenon, forecasting weather and warn about the extreme events. Government along with the local administration should create a think tank in order to improve land management.

Author details

Zara Tariq and Shakeel Mahmood*
Department of Geography, Government College University, Lahore, Pakistan

*Address all correspondence to: shakeelmahmoodkhan@gmail.com

References

[1] Arekhi S, Niazi Y, Kalteh AM. Soil erosion and sediment yield modeling using RS and GIS techniques: A case study, Iran. Arabian Journal of Geosciences. 2012;**5**(2):285-296

[2] Arekhi S, Shabani A, Rostamizad G. Application of the modified universal soil loss equation (MUSLE) in prediction of sediment yield (case study: Kengir watershed, Iran). Arabian Journal of Geosciences. 2012;**5**(6):1259-1267

[3] Hlaing KT, Haruyama S, Aye MM. Using GIS-based distributed soil loss modeling and morphometric analysis to prioritize watershed for soil conservation in Bago river basin of lower Myanmar. Frontiers of Earth Science in China. 2008;**2**(4):465-478

[4] Pradeep GS, Krishnan MV, Vijith H. Identification of critical soil erosion prone areas and annual average soil loss in an upland agricultural watershed of Western Ghats, using analytical hierarchy process (AHP) and RUSLE techniques. Arabian Journal of Geosciences. 2015;**8**(6):3697-3711

[5] Sharda VN, Mandai D, Ojasvi PR. Identification of soil erosion risk areas for conservation planning in different states of India. Journal of Environmental Biology. 2013;**34**(2):219

[6] Zakerinejad R, Maerker M. An integrated assessment of soil erosion dynamics with special emphasis on gully erosion in the Mazayjan basin, southwestern Iran. Natural Hazards. 2015;**79**(1):25-50

[7] Gayen A, Pourghasemi HR, Saha S, Keesstra S, Bai S. Gully erosion susceptibility assessment and management of hazard-prone areas in India using different machine learning algorithms. Science of the Total Environment. 2019;**668**:124-138

[8] Das B, Paul A, Bordoloi R, Tripathi OP, Pandey PK. Soil erosion risk assessment of hilly terrain through integrated approach of RUSLE and geospatial technology: A case study of Tirap District, Arunachal Pradesh. Modeling Earth Systems and Environment. 2018;**4**(1):373-381

[9] Zia-ur-Rehman M, Murtaza G, Qayyum MF, Ullah S, Rizwan M, Ali S, et al. Degraded soils: Origin, types and management. In: Soil Science: Agricultural and Environmental Prospectives. Cham: Springer; 2016. pp. 23-65

[10] Khan A, Rahman AU, Mahmood S. Spatial estimation of soil erosion risk using RUSLE model in District Swat Eastern Hindu Kush, Pakistan. Journal of Water and Climate Change. 2023

[11] Smith SJ, Williams JR, Menzel RG, Coleman GA. Prediction of sediment yield from southern plains grasslands with the modified universal soil loss equation. Rangeland Ecology & Management/ Journal of Range Management Archives. 1984;**37**(4):295-297

[12] Mitasova H, Hofierka J, Zlocha M, Iverson LR. Modelling topographic potential for erosion and deposition using GIS. International Journal of Geographical Information Systems. 1996;**10**(5):629-641

[13] Renard KG, Foster GR, Weesies GA, Porter JP. RUSLE: Revised universal soil loss equation. Journal of Soil and Water Conservation. 1991;**46**(1):30-33

[14] Shaw R. Floods in the Hindu Kush region: Causes and socio-economic

aspects. Mountain Hazards and Disaster Risk Reduction. 2015:33-52

[15] Scott CA, Zhang F, Mukherji A, Immerzeel W, Mustafa D, Bharati L. Water in the Hindu Kush Himalaya. In: The Hindu Kush Himalaya Assessment: Mountains, Climate Change, Sustainability and People. Cham: Springer; 2019. pp. 257-299

[16] Mahmood S, Rahman A. Flash flood susceptibility modelling using geo-morphometric and hydrological approaches in Panjkora Basin, Eastern Hindu Kush, Pakistan. Environmental Earth Sciences. 2019; **78**(1):43-58

[17] Mahmood S, Rahman AU, Shaw R. Spatial appraisal of flood risk assessment and evaluation using integrated hydro-probabilistic approach in Panjkora River Basin, Pakistan. Environmental Monitoring and Assessment. 2019; **191**(9):573

[18] Wischmeier WH, Smith DD. Predicting Rainfall Erosion Losses: A Guide to Conservation Planning. Vol. 537. Department of Agriculture, Science and Education Administration; 1978. pp. 1-75

[19] Maqsoom A, Aslam B, Hassan U, Kazmi ZA, Sodangi M, Tufail RF, et al. Geospatial assessment of soil erosion intensity and sediment yield using the revised universal soil loss equation (RUSLE) model. ISPRS International Journal of Geo-Information. 2020; **9**(6):356

[20] Ullah W, Nafees M, Khurshid M, Nihei T. Assessing farmers' perspectives on climate change for effective farm-level adaptation measures in Khyber Pakhtunkhwa, Pakistan. Environmental Monitoring and Assessment. 2019; **191**(9):1-18

[21] Ganasri BP, Ramesh H. Assessment of soil erosion by RUSLE model using remote sensing and GIS-A case study of Nethravathi Basin. Geoscience Frontiers. 2016;**7**(6):953-961

[22] Ullah S, Ali A, Iqbal M, Javid M, Imran M. Geospatial assessment of soil erosion intensity and sediment yield: A case study of Potohar region, Pakistan. Environmental Earth Sciences. 2018; **77**(19):1-13

[23] Sheikh AH, Palria S, Alam A. Integration of GIS and universal soil loss equation (USLE) for soil loss estimation in a Himalayan watershed. Recent Research in Science and Technology. 2011;**3**(3):51-57

[24] Osinski E, Meier U, Büchs W, Weickel J, Matzdorf B. Application of biotic indicators for evaluation of sustainable land use—Current procedures and future developments. Agriculture, Ecosystems and Environment. 2003;**98**(1-3):407-421

Chapter 3

Spatial Soil Loss Assessment Using USLE in Lake Ol Bolossat Catchment

John Mwangi, Charles K. Gachene, Stephen M. Mureithi and Boniface Kiteme

Abstract

Erosion by water is one of the most common types of soil degradation which occurs in all climatic regions and is widely considered to be a serious threat to the long-term viability of agriculture in many parts of the world. Lake Ol Bolossat in Nyandarua County, Kenya, is a high altitude lake that was formed on Rift Valley escarpment and faces the challenge of siltation due to increased soil erosion. Over the last few decades, the lake has been encroached and lake area has been overgrazed reducing the vegetation cover around the Lake. An assessment of spatial soil erosion loss was conducted using USLE model and GIS which showed that most parts of the Lake catchment have soil loss beyond tolerable levels of nine tons per year. The soil erosion range was between zero and 22, 525.5 tons per year. The land uses that were more vulnerable to soil loss are croplands, grazing lands with sparse vegetation and barelands which had soil loss ranges of 10–50 tons, 100 to 1000 tons and 500 to 22,525.5 tons per year respectively. The study recommended for immediate interventions by policy makers, researchers and development partners in curbing the soil loss problem.

Keywords: USLE, spatial soil erosion, Nyandarua, soil degradation, modelling, GIS

1. Introduction

Soil is a prerequisite for food production [1] especially at the wake of rapidly increasing world population placing a high demand for food resulting to agricultural intensification globally. However, food productivity is fast declining due to increased soil degradation [2]. Often, rapid land use transformation, for example, conversion of forests to agricultural land results to soil degradation which normally takes four main forms: water erosion; wind erosion; chemical degradation; and physical degradation. Each form of soil degradation, occurring both individually and in combination with the other forms, can result in the loss or damage to key ecosystem functions and processes [3–5]. Among the four forms of soil degradation, erosion by water is the most common which occurs in all agro-climatic zones and is widely considered to be a serious threat to the long-term agricultural production in many parts of the world [6–8]. It is a primary agent of soil degradation [9], affecting 1094 million ha, or roughly 56% of the land experiencing human-induced degradation worldwide [2, 10, 11]. Soil erosion has also been recognised to be the major non-point pollution source in many areas, which causes a large amount of damage every year [6, 12].

Soil erosion, considered as the most widespread form of soil degradation, has greatly affected agricultural production globally and in particular, Sub Saharan Africa [2, 6, 11, 13–16]. One-third of arable land globally has been estimated to have been lost over the last four decades due to soil erosion [2, 11] at a rate of over 10 million hectares per year [4]. Soil erosion by water dislodges soil particles from the surface due to impact of rain drops [14, 17, 18] which generate enough power to carry the particles far away causing sedimentation of water bodies [16, 19] and other environmental hazards [4, 20, 21]. These negative environmental effects usually deprive soil its capacity to perform its functions and the links between soil and other ecosystem components [2].

Soil erosion by water is sometimes considered to be a purely natural process caused by rainfall and water flow [4]; however human activities greatly aggravate the erosion through alteration of land cover and disturbance of soil structure through cultivation [2, 5, 11].

The main physical parameters influencing the intensity of erosion processes are climate regime, soil characteristics, topography and vegetation. Apart from these physical parameters, man and man-induced land use often have a significant influence on erosion intensity. When the protection of the natural vegetation cover is replaced by a temporary cover such as during the cultivation cycle, the erosion intensity might increase significantly [22]. Wischmeier and Smith [23] noted six (6) factors that affect soil erosion which include rainfall erosivity, soil erodibility, slope length and angle, crop management and conservation practices.

Assessment of soil erosion remains critical as it enables different players to formulate mitigation measures [4]. Various methods employed in assessing soil erosion - either by water or wind - vary depending on the causes, magnitude of erosion and on the scale of assessment as well as the applicability of the methods in different environments [8, 21, 24]. The most common methods of assessment are: expert opinions, land users' opinions [25, 26], field monitoring, observations and measurement, modelling [27, 28], estimates of productivity changes and remote sensing [29, 30].

The soil loss prediction models include Universal Soil Loss Equation (USLE), Revised Universal Soil Loss Equation (RUSLE), and Soil Loss Estimation Model for Southern Africa (SLEMSA) and can be helpful in designing sustainable land use practices in order to curb soil erosion menace [6, 11, 20]. These methods differ in their use and application depending on various factors such as the intentions for use; characteristics of study area; data requirement and availability; validity and reliability of the method [16, 21, 26]. These mathematical models are continually being improved and scientists from many countries have adopted them to meet the requirements of their local conditions [24, 31].

The USLE model is widely used to predict average annual rate of soil erosion based on rainfall intensity, type of soil, slope steepness, crop and soil management practices [7, 8, 16, 21, 24, 28, 31]. It is based on the product of rainfall-runoff erosivity (R), soil erodibility (K), slope length and steepness (LS), surface cover and management (C) and supporting conservation practices (P) [7, 21]. Nontananandh and Changnoi [10] noted that USLE requires relatively simple data and it is compatible with a geographic information system (GIS). Several studies point out that USLE remains the best available model that has been tested in virtually all environments of the world in spite of having been criticised and accused of giving erroneous results [17, 19, 28, 31]. The main criticism of the USLE has emanated from people having applied the model in environments in which it was not intended to be used [31]. This study aimed at assessing spatial soil loss using USLE model and GIS in Lake Ol Bolossat Catchment.

2. Materials and methods

2.1 Study area

Lake Ol Bolossat catchment is located on latitude 00 09′S and longitude 360 26′E in Nyandarua County, Central Province of Kenya. The lake with an area of 43.3 km^2 lies at an average altitude of 2340 m above sea level in a wedge shaped rift valley floor sloping eastwards and northwards [32] and forms the headwaters of Ewaso Narok River which is the major tributary of Ewaso Ng'iro North River (see **Figure 1**).

The region enjoys favourable climate for most periods of the year, with temperatures ranging between 10° and 28°C. The climate is sub-humid and is strongly influenced by local topography of the surrounding central highlands with mean

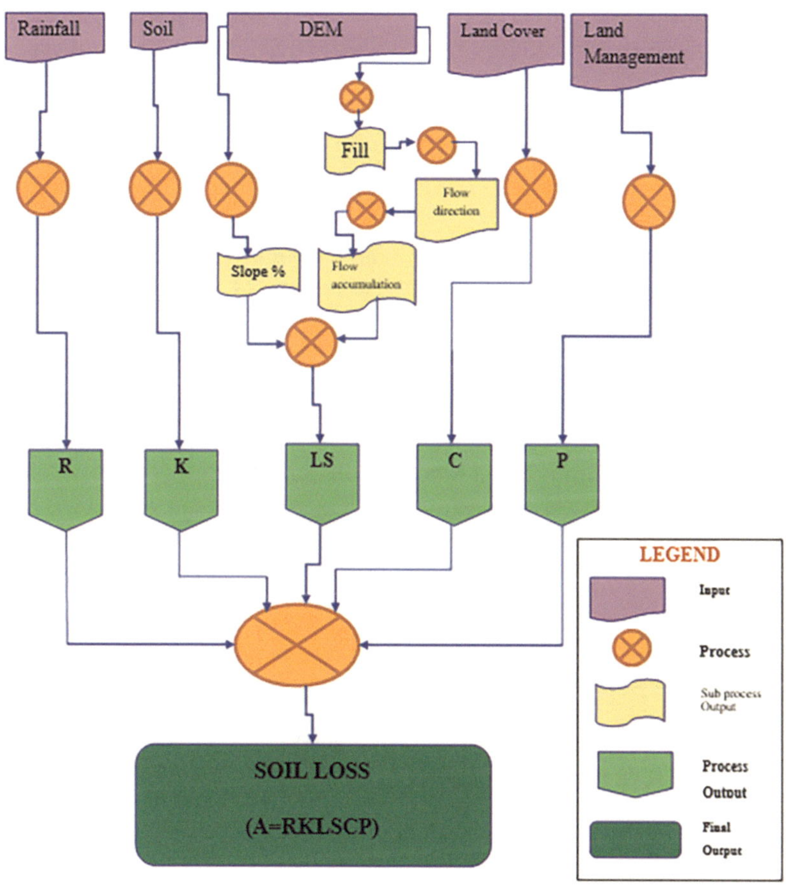

Figure 1.
Flow chart analysis of soil loss.

annual rainfall of 980 mm and increases southwards and westwards. Rainfall is bimodal, with long peaks between April and June and the shorter peaks between October and November [32].

The areas is dominated by small holder mixed farmers who grow crops and rear livestock on parcels of land ranging from 1 to 4 hectares. Nearly 60% of the families own less than 2 ha of land. Since they have free access to pasture around the lake, most of them own more livestock than their 2-ha plots can support. The human population density in the lake basin and the lake's watershed is approximately 202 per km^2 [33].

2.2 Data sources

Different data sources were referred to analyse the soil loss in the study area. A digital elevation model (DEM) with 90-meter resolution developed by NASA was used to calculate the slope length and slope gradient of the study area. The land cover classification map for 2014 was used for the analysis of crop management factor (C-value) while a soil map made by Centre for Training and Integrated Research in ASAL Development (CETRAD) was used in the analysis of soil erodibility factor (K-value).

Analysis of soil erosivity factor (R-value) was derived from annual rainfall data for different rainfall stations in the catchment which was obtained from CETRAD database. Conservation practices factor (P-value) was derived from land use types aligned to specified slope of the study area. The estimation of soil loss was then done by map overlays, pixel by pixel which enabled accurate multiplication of USLE parameters.

2.3 Methodology

The universal soil loss equation (USLE), developed by [23], was employed to assess the amount of soil loss in Lake Ol Bolossat Catchment. The USLE was applied in GIS based on the flow chart shown in **Figure 2**.

Mathematically the equation is denoted as:

$$A(tons/ha/yr) = RXKXLXSXCXP \quad (1)$$

Where A is the mean annual soil loss, R is the rainfall erosivity factor, K is the soil erodibility factor, L is the slope length factor, S is the slope steepness factor, C is the crop management factor and P is the erosion control practice or land management factor. The analysis of each process factors of USLE was derived procedurally as illustrated in **Figure 2**.

2.3.1 Rainfall erosivity factor (R)

There are three equations which have been used to derive R-factor in different parts of the world [30] namely;

$$R = 9.28XP - 8838 \quad (2)$$

Mean annual erosivity (KE > 25) where P is mean annual precipitation.

$$R = 0.276XPX130 \quad (3)$$

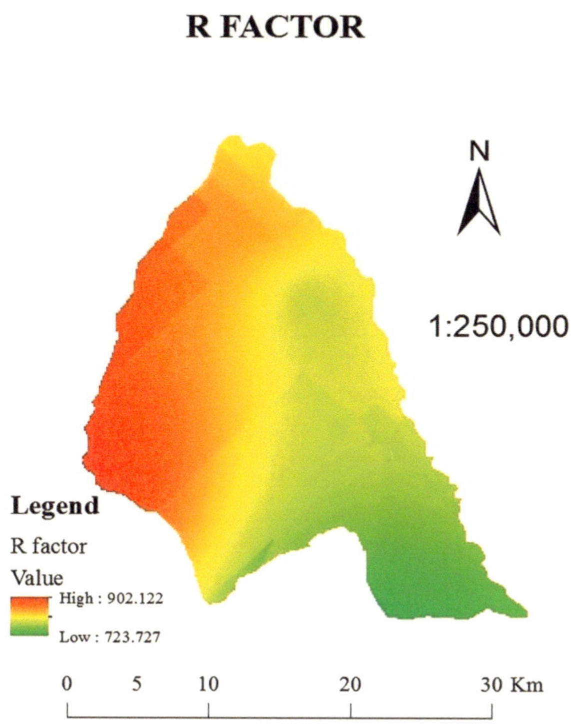

Figure 2.
Rainfall erosivity factor.

Mean annual EI30, where P is mean annual precipitation.

$$R = 0.5XP \text{ (in US unit) and } R = 0.5XPX1.73 \text{ (in Metric unit)} \quad (4)$$

The study noted that the Eq. (2) is applicable in Peninsular Malaysia while the Eq. (3) requires I30 factor which is difficult to calculate. Eq. (4) is applicable in humid and sub-humid areas with mean annual rainfall of between 900 mm and 1700 mm [30]. This equation Eq. (4) was applied in this study where it was integrated into Arc GIS 10.3 software to derive R-factor. The rainfall data was obtained from CETRAD database. Average annual rainfall for at least twenty years was computed for ten weather stations in the catchment with respective rainfall erosivity factor using Arc GIS 10.3 [e.g., **Table 1** and **Figure 3**).

2.3.2 Soil erodibility factor (K)

Bizuwerk et al. [30] defines Soil Erodibility Factor (K) as mean annual rainfall soil loss per unit of R for a standard condition of bare soil, recently tilled up-and-down with slope with no conservation practices and on a slope of 50 and 22 m length. Hellden [34] in [30] developed a USLE for humid and sub-humid highlands condition by adapting different sources and proposed the K values of the soil based on their colour. This soil classification was adopted for the study and modified according to four soil types found in the area. Soil map was obtained from CETRAD database.

Name of station	Location		Mean annual rainfall (P)	Erosivity (R)
	Longitude	Latitude		
Shamata Gate	−0.19457	36.52173	1121.72	970.29
Ndaragwa Forest Station	−0.06568	36.53005	921.82	797.37
Nyahururu Meteorological Station	0.02973	36.36449	1048.13	906.63
Ol bolossat Forest Station	−0.05476	36.33653	972.18	840.94
Ol Joro-Orok KARI	−0.01097	36.38219	765.5	662.16
South Marmanet Forest Station	0.04496	36.37365	970.75	839.70
Mirangine Chief's Camp	−0.17906	36.24434	1171	1012.915
Kangui Secondary School	−0.05081	36.39671	897	775.905
Ol Kalou Railway station	−0.27275	36.37752	694	600.31
Rumuruti MOW office	0.26748	36.54844	703.2	608.268

The rainfall stations were selected based on their proximity to the study area. Where there was no rainfall station existing near the study area, another station was selected and then extrapolation of data was carefully done in order to have the most representative rainfall data.

Table 1.
Computation of rainfall erosivity factor (R Factor) for the study area.

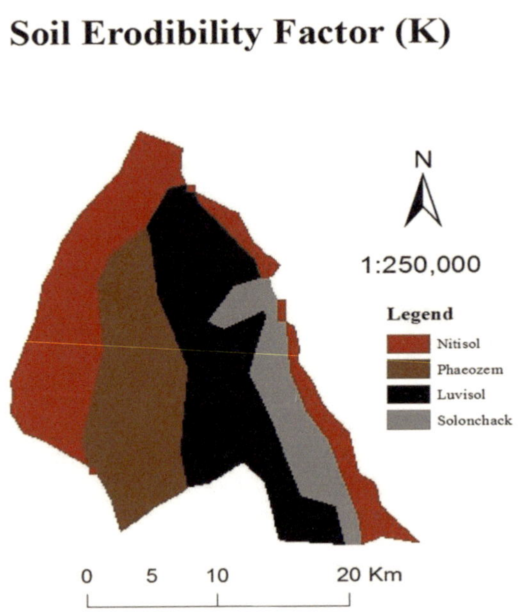

Figure 3.
Soil erodibility factor.

However, the soil data were in their geomorphologic names but not their colour and hence an attempt was made to match the soil names with their colour referring to World Reference Bureau (WRB) classification. The k value was then computed using Arc GIS 10.3 and results presented in raster format as shown in **Figure 4**.

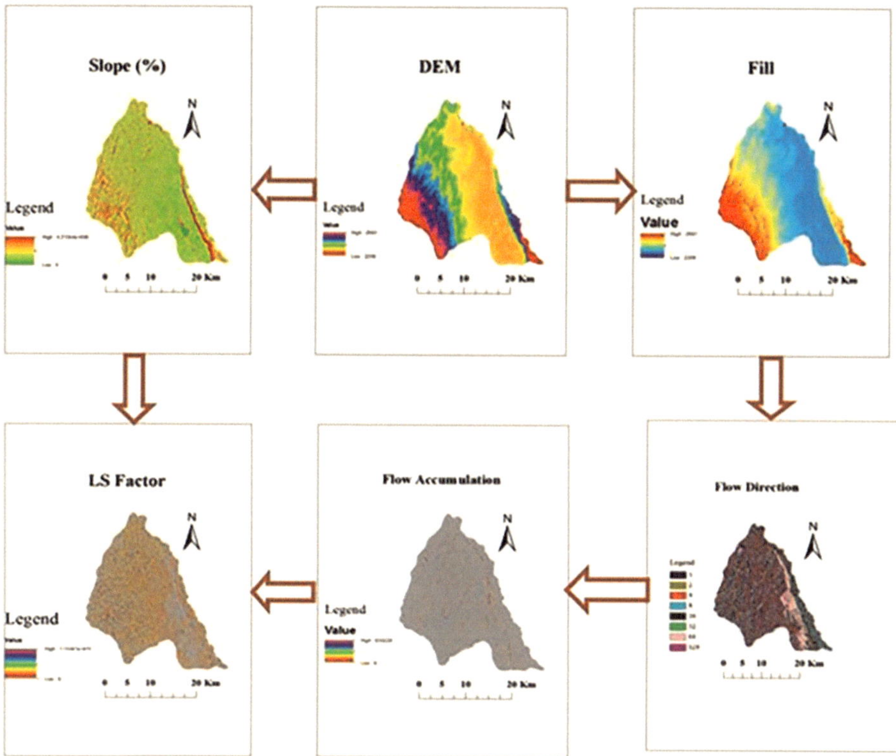

Figure 4.
LS factor.

2.3.3 Slope length and slope steepness factors (LS)

Wischmeier and Smith [23] noted that slope length and slope steepness can be used in a single index to express the ratio of soil loss using the following equation.

$$LS = (X/22.1)^m \left(0.065 + 0.045S + 0.0065S^2\right) \tag{5}$$

Where X = slope length (m) and S = slope gradient (%).

The values of X and S were derived from DEM. To calculate the X value, Flow Accumulation was derived from the DEM after conducting FILL and Flow Direction processes in Arc GIS 10.3

$$\text{X} = (\text{Flow accumulation}^* \ \text{Cell value})$$

By substituting X value, LS equation was:

$$\text{LS} = (\text{Flow accumulation} * \text{Cell value}/22.1)^{\text{m}} \left(0.065 + 0.045\ \text{S} + 0.0065\ \text{S}^2\right)$$

Moreover slope (%) was directly calculated from the DEM using the same software. The value of m varies from 0.2–0.5 depending on the slope as shown in **Table 2** [23]. The result of LS factor analysis is shown in the **Figure 5**.

m-value	Slope (%)
0.5	> 5
0.4	3–5
0.3	1–3
0.2	<1

The slope for the study area ranged from less than 1%–5% and therefore the m-value (slope length) ranged from 0.2 to 0.5 (dimensionless).
Source: Bizuwerk et al. [30].

Table 2.
m-value.

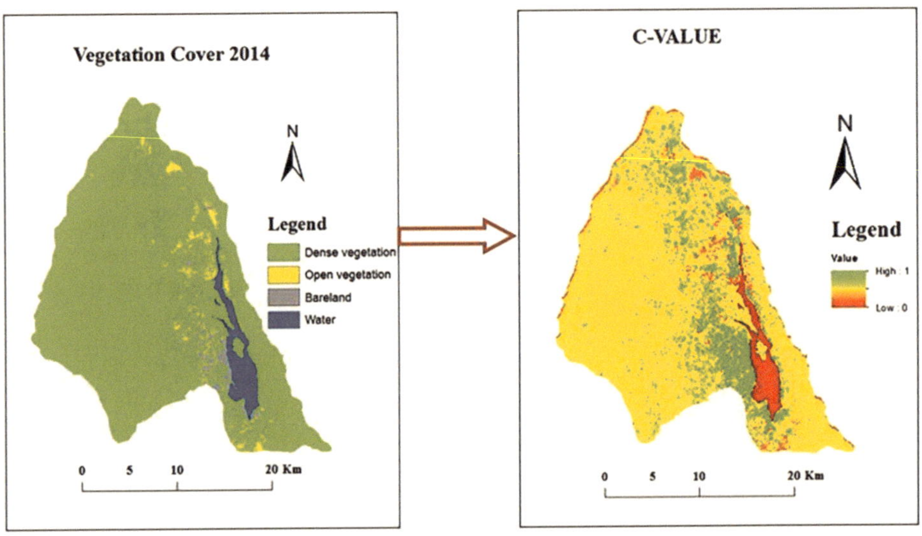

Figure 5.
Crop management factor.

2.3.4 Crop management factor (C)

The crop management factor which represents the ratio of soil loss under a given crop to that of the base soil [30] is perhaps the most important USLE factor because it represents conditions that could be managed most easily to reduce erosion [10]. The land use map was developed and used for analysing the c-value. The coverage was changed to grid where a corresponding c-value was assigned to each land cover class using reclass method in ArcGIS 10.3 and results presented as shown in **Figure 6**.

2.3.5 Erosion management practice factor (P-value)

The support practice factor, P, is a soil loss ratio for a specific support practice to the corresponding soil loss with up-and-down slope tillage [10]. The P-value ranges from 0 to 1 depending on the soil management activities applied in the specific plot of

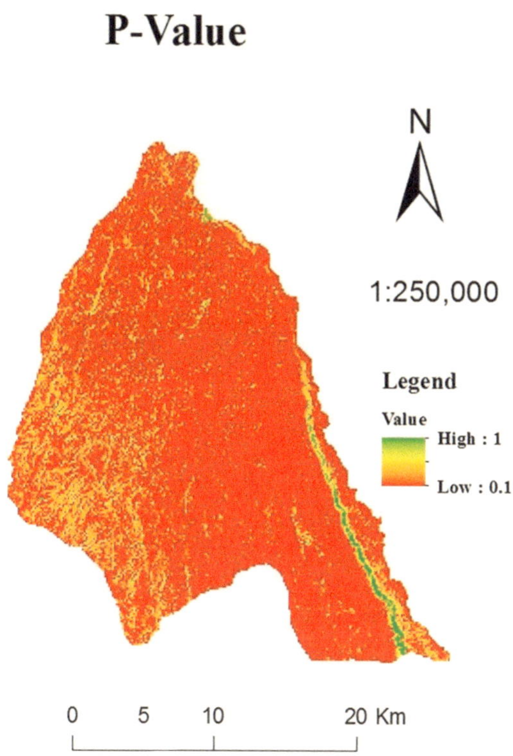

Figure 6.
Erosion management factor.

Land use type	Slope (%)	P-factor
Agricultural land	0–5	0.1
	5–10	0.12
	10–20	0.14
	20–30	0.19
	30–50	0.25
	50–100	0.33
Other land	All	1.00

Source: Bizuwerk et al. [30].

Table 3.
P-value.

land. These management activities highly depend on the slope of the area [30]. Wischmeier and Smith [23] calculated the P-value by delineating the land into two major land uses, agricultural land and other land. The agricultural land sub-divided into six classes based on the slope percent to assign different P-value as shown in **Table 3**. The study applied this same technique to assign the P-value of the catchment. The results were computed and presented in raster format as shown in **Figure 7**.

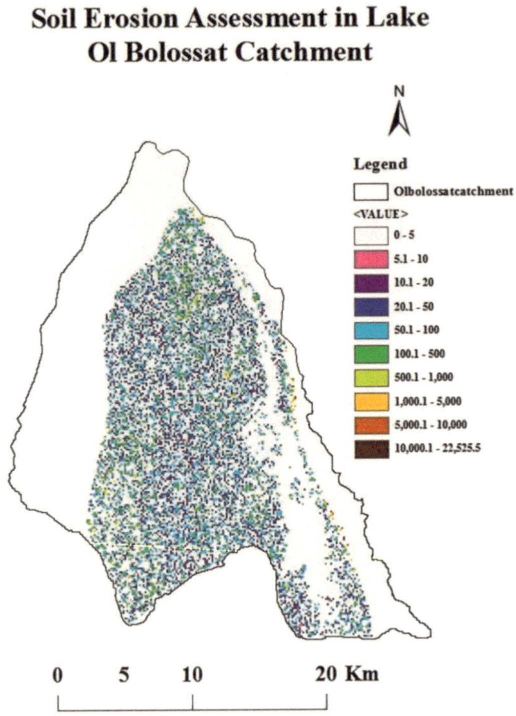

Figure 7.
Estimated soil loss.

3. Results

The results of this study show that the amount of soil loss in Lake Ol Bolossat catchment ranges between zero (0) and 22, 525.5 tons per year from 57, 800 hectares as shown in **Figure 8**. This implies that the mean annual soil loss is approximately 0.389 tons per hectare per year. However, averaging the soil loss would suggest that the study area is experiencing low soil erosion which is much lower than tolerable levels of 9 ton per hectare [35].

The study revealed that areas covered with dense vegetation i.e. along Ndudori Tumaini escarpment and South Marmanet forest have little or no soil loss while areas with sparse vegetation and Bareland have high soil loss. It is also noted from the study that areas dominated by agricultural activities and areas around the Lake have experienced high soil loss. In addition, barelands and areas with sparse vegetation have high soil erosion rate. Most parts of the study area did not have soil conservation measures as indicated by low value of P factor in **Figure 7** and this could be another possible reason why these areas have high erosion rate.

4. Discussion

Rainfall erosivity factor (R) plays a vital role in soil loss. The catchment under study has R-value of between 723.7 and 902.1 (**Table 1** and **Figure 3**). High R-factor is usually associated with soil loss due to high kinetic energy of rain drops that dislodges

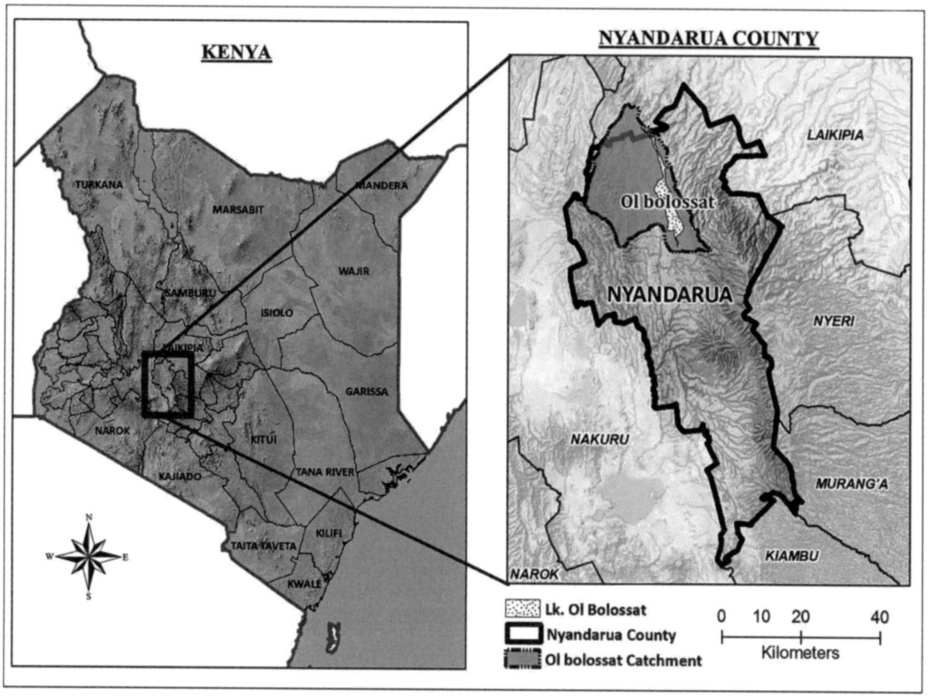

Figure 8.
Study area.

and disintegrates soil particles that are easily carried away by surface runoff [3, 36, 37]. However, other factors such as vegetation cover and erosion management factors also determine the rate of soil erosion [22, 36]. According to the study areas with high R-values and high vegetation cover such as western strip of study area have little soil loss (less than 5 tons/year). In comparison, areas with moderate R-value (about 723.7) and little or no vegetation cover experience high soil loss indicating that C-factor plays a major role in controlling soil erosion. In the view of this, it can be concluded that there is no soil erosion factor acts in isolation but all USLE factors are interrelated.

Vegetation cover is an important determinant of soil erosion. Vegetation intercepts the rain drops reducing their impact on soil particles. In addition, vegetation cover slows down the surface runoff allowing for more for water to infiltrate and consequently reducing the scouring ability of the runoff [29, 36, 37].

The type of soil in an area determines the soil erodibility value. The study area has four dominant soil types i.e. nitisol, phaeozem, luvisol and solonchack which have different erodibility values (**Table 4**). The results of the study show that nitisol is erodible even though they have a high erodibility value (0.25) because these soils are deep and supports dense vegetation. Solonchacks are highly erodible (K-value of 0.35) and areas dominated by this type of soil have high erosion rate. Most of these areas are around the Lake and have a soil loss of over ten tons per hectare per year which is above tolerable values of 9 tons per hectare per year [35, 37]. Luvisols and phaeozems are relatively erodible and areas dominated by these soils have high rate of soil erosion. In addition to soil types, overgrazing also contributes to high soil erosion rate

Soil colour	Reddish brown	Brown	Black	Grey
Soil Type	Nitisol	Phaeozem	Luvisol	Solonchack
K factor	0.25	0.2	0.15	0.35

Modified from Bizuwerk et al. [30].

Table 4.
k-values.

especially around the Lake. The same observation was made by [35] when they were assessing erosion hazard in Upper Ewaso Ng'iro Basin of Kenya.

Slope length and slope angle (LS factors) also influence the rate of soil erosion. Steep slope especially Satima escarpment experience high rate of soil erosion. These sloppy areas have relatively high vegetation cover and erosion management practices as shown in **Figures 5** and **6** respectively, however the erosion rate is high possibly due to high erosivity factor.

The results of this study are based on use of empirical model, USLE. It is therefore recommended that other methods such as experimental plots be used so as to compare the results. The results of prediction models can be improved by carrying out semi-detailed or detailed soil surveys so as to enable a better way of computing erodibility factor. In addition, field assessment of soil and water conservation measures would help to accurately determine erosion management practices (P-factor). In addition, further research is recommended on assessment of soil erosion hot spots so as to formulate immediate soil conservation measures.

The study revealed that the spatial rate of soil loss in Lake Ol Bolossat catchment is between zero (0) and 22,575.5 tons per year. The rate of soil loss in most of the parts of the study area is alarming and requires mitigation measures for it is above the tolerable levels of 9 tons per hectare. The study also revealed that rainfall intensity, soil erodibility, slope length, slope angle, vegetation cover and erosion management are factors that contribute to soil erosion in Lake Ol Bolossat catchment. The spatial locations of high erosion rates are agricultural areas (croplands) and grazing areas around the Lake which are dominated by barelands and sparse vegetation.

Acknowledgements

The authors wish to acknowledge financial and platform support provided by Centre for Training and Integrated Research in ASAL Development (CETRAD) and supervisory support by department of Land Resource Management and Agricultural Technology (LARMAT), University of Nairobi. In addition, we wish to thank Madam Rahab Nyururu, Water Resources Authority (WRA), Rumuruti who was the lead research assistant and coordinated field data collection. Finally, we express our gratitude to the community in Lake Ol Bolossat catchment for continued support during the study.

Conflict of interest

The authors and other stakeholders involved in the study have no conflict of interest whatsoever.

Author details

John Mwangi[1,2]*, Charles K. Gachene[1], Stephen M. Mureithi[1] and Boniface Kiteme[2]

1 Department of Land Resources Management and Agricultural Technology, University of Nairobi, Nairobi, Kenya

2 Centre for Training and Integrated Research in ASAL Development, Nanyuki, Kenya

*Address all correspondence to: mwangijc2001@gmail.com

References

[1] Stenberg B, Raphael A, Rossel V, Mouazen A, Wetterlind J. Visible and near infrared spectroscopy in soil science. In: Sparks DL, editor. Advances in Agronomy. Vol. 107. Burlington: Academic Press; 2010. pp. 163-215. DOI: 10.1016/s0065-2113(10)07005-7

[2] Yang D, Kanae S, Oki T, Koike T, Musiake K. Global potential soil erosion with reference to land use and climate changes. Hydrological Processes. 2003; **17**:2913-2928. DOI: 10.1002/hyp.1441

[3] Stringer LC. Global Land and Soil Degradation: Challenges to Soil. Sustainability Research Institute, School of Earth and Environment. Leeds, West Yorkshire: University of Leeds; 2013

[4] Ganasri BP, Ramesh H. Assessment of soil erosion by RUSLE model using remote sensing and GIS-A case study of Nethravathi Basin. Geoscience Frontiers. 2016;**7**:953-961. DOI: 10.1016/j.gsf.2015.10.007

[5] Orchard SE, Stringer LC, Manyatsi AM. Farmer perceptions and responses to soil degradation in Swaziland. Land Degradation Development. 2017;**28**:46-56. DOI: 10.1002/ldr.2595

[6] Angima D, Stott D, O'Neill MK, Ong CK, Weesies GA. Soil erosion prediction using RUSLE for central Kenyan highland conditions. Agriculture, Ecosystems and Environment. 2003;**97**:295-308. DOI: 10.1016/S0167-8809(03)00011-2

[7] Cohen MJ, Shepherd KD, Walsh MG. Empirical reformulation of the universal soil loss equation for Erosion risk assessment in a tropical watershed. Geoderma. 2005;**124**:235-252. DOI: 10.1016/j.geoderma.2004.05.003

[8] Devatha CP, Deshpande V, Renukaprasad MS. Estimation of soil loss using USLE model for Kulhan watershed, Chattisgarh – A case study. Aquatic Procedia. 2015;**4**:1429-1436. DOI: 10.1016/j.aqpro.2015.02.185

[9] Ali SA, Hagos H. Estimation of soil erosion using USLE and GIS in Awassa catchment, Rift valley, Central Ethiopia. Geoderma Regional. 2016;**7**:159-166. DOI: 10.1016/j.geodrs.2016.03.005

[10] Nontananandh S, Changnoi B. Internet GIS, based on USLE Modeling, for assessment of soil Erosion in Songkhram watershed, Northeastern of Thailand. Kasetsart Journal (Nat. Sci.). 2012;**46**:272-282 (doi not available)

[11] Gelagay HS, Minale AS. Soil loss estimation using GIS and remote sensing techniques: A case of Koga watershed, Northwestern Ethiopia. International Soil and Water Conservation Research. 2016;**4**:126-136. DOI: 10.1016/j.iswcr.2016.01.002

[12] Breiby T. Assessment of Soil Erosion Risk within a Subwatershed Using GIS and RUSLE with a Comparative Analysis of the Use of STATSGO and SSURGO Soil Databases. Winona, MN: Saint Mary's University of Minnesota Central Services Press; 2006. p. 22. Retrieved on 5/20/2014 from http://www.gis.smumn.edu (doi not available)

[13] Pimentel D, Harvey C, Resosudarmo P, Sinclair K, Kurz D, McNair M, et al. Environmental and economic costs of soil Erosion and conservation benefits. Science, New Series. 1995;**267**(5201):1117-1123 accessed online (http://www.jstor.org/stable/2886079) on 20-09-2016

[14] Pimentel D. Soil erosion: A food and environmental threat. Environment,

Development and Sustainability. 2006;**8**: 119-137. DOI: 10.1007/s10668-005-1262-8

[15] Panagos P, Meusburger K, Ballabio C, Borrelli P, Alewell C. Soil erodibility in Europe: A high-resolution dataset based on LUCAS. Science of the Total Environment. 2014;**479–480**:189-200. DOI: 10.1016/j.scitotenv.2014.02.010

[16] Aiello A, Adamo M, Canora F. Remote sensing and GIS to assess soil erosion with RUSLE3D and USPED at river basin scale in southern Italy. Catena. 2015;**131**:174-185. DOI: 10.1016/j.catena.2015.04.003

[17] Rodriguez JLG, Suarez MCG. Methodology for estimating the topographic factor LS of RUSLE3D and USPED using GIS. Geomorphology. 2012;**175–176**:98-106. DOI: 10.1016/j.geomorph.2012.07.001

[18] Ashiagbor G, Forkuo EK, Laari P, Aabeyir R. Modeling soil erosion using RUSLE and GIS tools. International Journal of Remote Sensing & Geoscience (IJRSG). 2013;**2**(4):6-17 ISSN No: 2319-3484

[19] Stefano CD, Ferro V, Burguet M, Taguas EV. Testing the long term applicability of USLE-M equation at a olive orchard microcatchment in Spain. Catena. 2016;**147**:71-79. DOI: 10.1016/j.catena.2016.07.001

[20] Prasannakumar V, Vijith H, Abinod S, Geetha N. Estimation of soil erosion risk within a small mountainous sub-watershed in Kerala, India, using revised universal soil loss equation (RUSLE) and geo-information technology. Geoscience Frontiers. 2011;**3**(2):209-215. DOI: 10.1016/j.gsf.2011.11.003

[21] Mahalingam B, Malik MM, Vinay M. Assessment of soil Erosion using USLE technique: A case study of Mysore District, Karnataka, India. Journal of Remote Sensing & GIS. 2015;**6**(3) (doi not available)

[22] Kiruna. Assessment of Soil Degradation in Nigeria. Abuja, Nigeria: Federal Republic of Nigeria, Federal Department of Agricultural Land Resources; 1999

[23] Wischmeier H, Smith D. Predicting rainfall Erosion losses: A guide to conservation planning. In: Series: Agricultural Handbook No. 537. Washington DC: USDA; 1978

[24] Bagarello V, Stefano C, Ferro V, Pampalone V. Predicting maximum annual values of event soil loss by USLE-type models. Catena. 2017;**155**:10-19. DOI: 10.1016/j.catena.2017.03.002

[25] Symeonakis E, Drake N. Monitoring desertification and land degradation over sub-Saharan Africa. International Journal of Remote Sensing. 2004;**25**(3): 573-592. DOI: 10.1080/0143116031000095998

[26] Omuto C, Balinti Z, Alim M. A framework for National Assessment of land degradation in the dry lands: A case study of Somalia. Land Degradation and Development. 2011;**2011**:1-15. DOI: 10.1002/ldr.1151

[27] Nachtergaele FO. Land Degradation Assessment Indicators and the LADA project, Soil Conservation and Protection in Europe (Scape). In: Briefing papers of the second SCAPE workshop in Cinque Terre (IT, held from 13-15 April 2004). Italy: Land and Water Development Division, FAO; 2004. pp. 201-213

[28] Kapalanga TS. A Review of Land Degradation Assessment Methods. Reykjavík, Iceland: Land Restoration Training Programme Keldnaholt; 2008

[29] Brady C, Weil R. The Nature and Properties of Soil. 13th ed. New Jersey, USA: Prentice Hall; 2002

[30] Bizuwerk A, Taddese G, Getahun Y. Application of GIS for Modelling Soil Loss Rate in Awash River Basin. Ethiopia, Addis Ababa: International Livestock Research Institute (ILRI); 2003

[31] Pham TG, Degener J, Kappas M. Integrated universal soil loss equation (USLE) and geographical information system (GIS) for soil erosion estimation in a sap basin: Central Vietnam. International Soil and Water Conservation Research. 2018;**6**:99-110. DOI: 10.1016/j.iswcr.2018.01.001

[32] Zacharia M, Elias A, Jeremiah K, Simon M, Olang LO. Assessment of land cover changes in Lake Olbolosat region of the central Kenyan highlands using Landsat satellite imagery aided by indigenous knowledge. Journal of Biodiversity Management Forestry. 2013;**2**:2. DOI: 10.4172/2327-4417.1000107

[33] Wiesmann U, Kiteme B, Mwangi Z. Socio-Economic Atlas of Kenya: Depicting the National Population Census by County and Sub-location. KNBS, Nairobi. CETRAD, Nanyuki. CDE, Bern; 2014. p. 152

[34] Helden U. An Assessment of Woody Biomass, Community Forests, Land Use and Soil, Erosion in Ethiopia. Sweden: Lund University Press; 1987

[35] Mati BM, Morgan RP, Gichuki FN, Quinton JN, Brewer TR, Liniger HP. Assessment of erosion hazard with the USLE and GIS: A case study of the upper Ewaso Ng'iro north basin of Kenya. International Journal of Applied Earth Observation and Geoinformation. 2000;**2**(2):78-86. DOI: 10.1016/s0303-2434(00)85002-3

[36] Lal R. Soil degradation by erosion. Land Degradation and Development. 2001;**12**:519-539. DOI: 10.1002/ldr.472

[37] Kiage L. Perspectives on the assumed causes of land degradation in the rangelands of Sub Saharan Africa. Progress in Physical Geography. 2013;**37**:664-684. DOI: 10.1177/0309133313492543

Chapter 4

Modeling of Soil Sensitivity to Erosion Using the Analytic Hierarchical Process: A Study of Menoua Mountain Watershed, West-Cameroon

Gabriel Nanfack and Moye Eric Kongso

Abstract

The Bamboutos Mountains experience a persistent deterioration of their natural environment, which is evidenced by the ongoing loss of vegetation and growing instability of the ecosystem. As such, several soil restoration projects have been put in place to restore this mountain ecosystem and maintain its agricultural potential. This article goes in-line with this premise by studying the sensitivity of soils to water erosion in a watershed where agriculture is the main form of land use. The objective of the study is to examine various aspects of the study area, including its topography, lithology, hydrology, climate, and land use, in order to adopt a multi-criteria approach that involves intersecting these factors related to soil vulnerability to erosion using GIS. Results showed that the Menoua watershed is characterized by very steep slope classes (60% of the area occupied by slopes greater than 50°), with agricultural land alone covering approximately 49% of the watershed or almost half of the available space. The map of soil sensitivity to erosion shows that areas most sensitive to erosion (42%) generally coincide with the sloping land cultivated on lateritic soils in the northern part of the basin. Very strong and strong sensitivity to erosion represents 8.82%. The basin is therefore a geographical area at risk of erosion. Adopting no-tillage farming technique and the agroforestry can reduce sensitivity to erosion and ensure sustainable management of mountains.

Keywords: sensitivity to erosion, agriculture, AHP, GIS, Menoua watershed

1. Introduction

The Menoua watershed is a hydrosystem where agriculture is the main form of land use [1]. The combined effect of ancestral farming practices, which often do not take soil conservation into account (slash-and-burn method, absence of fallows, ridging along steep slope) [2], high rainfall (2000 mm/year), relief, soil type (ferralitic soils), and the regression of the plant cover promotes water erosion. This leads to soil degradation and a drop in agricultural production potential. This phenomenon creates imbalances and damage in the production basins: degradation of the topsoil with loss of fertilizing elements upstream and excessive deposits of alluvium downstream [3, 4].

As such, the need to achieve risk reduction by identifying potential sites with high erosive sensitivity is the major objective of this study. It will be a question of researching for this intra-mountain space, the natural and/or anthropogenic predispositions that characterize the sites prone to this phenomenon of soil erosion. It is therefore important to identify these areas by taking into account the biophysical and environmental realities, which predominate in the watershed in order to improve the sustainability of production systems. Therefore, in order to map the sensitivity of soils to erosion in the Menoua watershed, an approach based on modeling, which makes it possible to intersect factors taken into account for the mapping of soil sensitivity to erosion using GIS was chosen. This chapter mainly deals with the methodology for mapping the sensitivity of soils to erosion. It is divided into four main parts. The first part presents the geographical framework of the study area, the second briefly details the methodology used while results and discussion are presented in the third and fourth parts, respectively.

2. Materials and methods

2.1 The study area

With a superficial area of 68,200 hectares, the Menoua watershed is part of the Western Highlands of Cameroon. It is located on the southern slope of the Bamboutos Mountains between latitude 5° and 6° North of the equator and 9° and 10° of the Greenwich Meridian. Its average altitude varies between 700 m downstream and 2270 m upstream (**Figure 1**). It has a compactness index of around 2.107 [5], which reduces the response time of the basin to precipitation and promotes erosion. It receives an average annual rainfall amount of 1900 mm [6], characterized by both spatiotemporal irregularity and high intensity [7]. It has an irregular slope system that exceeds 15° on nearly 70% of the basin [8] and a reduced vegetation cover that now represents 28.35% of the catchment area [1].

Like in all of the Western Highlands, geological formations are mainly made up of products of volcanic eruptions (rhyolitic, trachytic, basaltic flows, and ignimbritic projections) of varying ages ranging from the tertiary to the current period [9]. These rocks consist of minerals with a fairly high degree of alterability [10]. The drainage density is high showing an impermeable lithology that favors surface runoff. The hydrographic network is of order 5, and therefore very active during intense and concentrated rains. However, it is a densely populated environment where agriculture is the main economic activity [11, 12]. Anthropogenic interventions remain the major process that directly influences the acceleration of water erosion processes on the slopes.

2.2 Data and materials

The analysis of soil sensitivity of the Menoua watershed was carried out on the basis of a set of data such as the Landsat 8 satellite image of 30 m spatial resolution h ttps://www.earthexplorer.org; the Bafoussam 1c and 1d topographic maps of 1/50,000

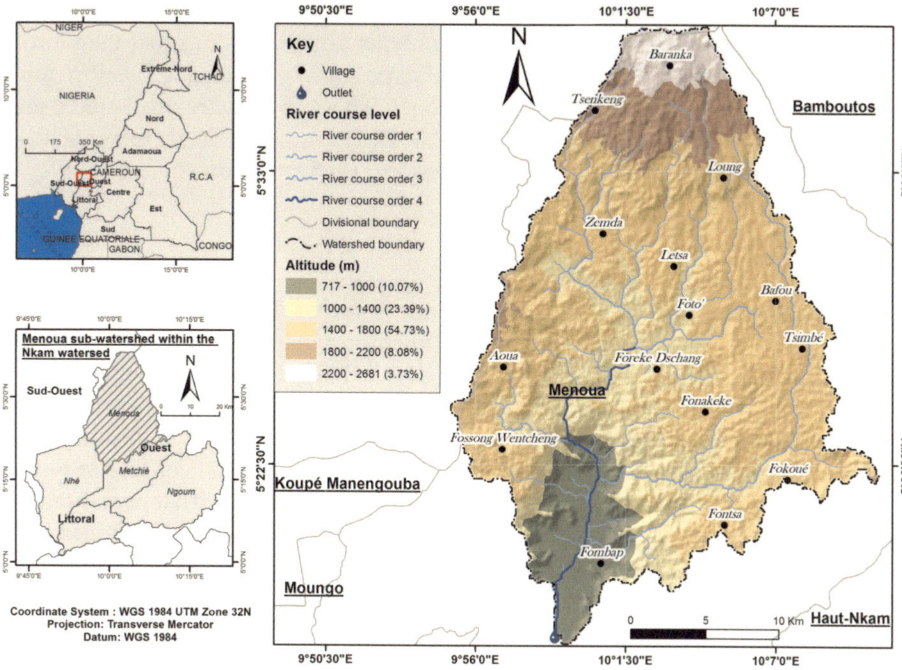

Figure 1.
Presentation and localization of the study zone.

provided by the National Institute of Cartography; the soil map extracted from the morpho-pedological map of West Cameroon at 1/100,000, and data collected using a navigation GPS during fieldwork on the southern slopes of the Bamboutos Mountains in December 2021. This data collection have enabled the establishment of a Spatial Reference database. ArcGIS 10.3 and Envi 5.2 software were used for data processing.

2.3 Methods

The hierarchical multi-criteria analysis applied in a Geographic Information System for the assessment of soil erosion hazard is increasingly used nowadays. Literature is abundant and has revealed that anthropogenic factors contribute to soil erosion [13–15]. The evaluation of soil sensitivity to erosion in the Menoua intra-mountain watershed made use of the Hierarchical Analytical Process (PAH) developed by Saaty [16]. Multi-criteria spatial evaluation methods are generally made up of six stages: the entry criteria, the hierarchical structuring of the criteria, the development of binary combinations, the determination of the value and proper vector, the study of consistency of judgment, and finally an aggregation structure, leading to a final relevance map [17].

2.3.1 Elaboration of the erosion sensitivity

The entry criteria are factors considered included in the decision-making process as having a major influence on soil erosion, and which can be characterized by their respective attributes. This stage aims to characterize the drainage basin from the topographical, lithological, land use, and climatic perspectives. In this study, the sensitivity of soils to erosion is assessed by applying GIS techniques that take into account the operations of extraction, reclassification, geo-referencing, vectorization, and rasterization based on multi-criteria analysis. These factors are evaluated using the spatial analysis capabilities of ArcMap 10.3. Each evaluation leads to a map representing the entire basin, and its suitability for the factor considered.

2.3.1.1 Topography

The topographic factor that reveals the slope is one of the first criteria for conditioning runoff. The slope greatly influences the importance of erosion by its gravitational action and provides its erosive energy to water. Thus, the action of erosion increases strongly with slope. Slope steepness acts directly on runoff velocity. When it increases, the kinetic energy of runoff increases and accelerates the transportation of solid objects downwards, increasing the impact of the ablation. According to Refs. [18, 19], the transportation of particles increases exponentially with a percentage inclination of the slope. Topography shows the inclination of relief and the length of the slope. These criteria are very important and directly influence the phenomenon of erosion because it conditions the speed of runoff. In the context of this study, we will retain only the inclination of slope, through its preponderance in the topographic parameter. To this effect, a topographic map of Bafoussam 1c at 1/50,000 provided by the National Institute of Cartography made it possible to digitalize the contour lines. ArcGIS 10.3 made it possible to establish in a precise manner, the interpolation of side points resulting from the contour lines on the slopes map of the basin. These slopes were ordered into five classes (**Table 1**).

2.3.1.2 Land use

The erosion process is closely linked to the mode of land use, which largely contributes either to its aggravation or attenuation. Land use determines the degree of soil protection. This factor is a measure of the relative effectiveness of soil and crop management systems in preventing or reducing soil loss [20]. In terms of soil protection, vegetation is essential because foliage, tree trunks, and roots constitute obstacles that slow down the speed of runoff through the phenomenon of "stem-flow." It also protects against the phenomenon of rain splashing, prolongs the permeability of soils, and reduces the volume of runoff. The degradation of woodlands leads to a profound disturbance in the hydrological regime of the basin and exposes soils, leading to

Pente (°)	Surface (%)	Influence on erosion
≤ 5	26.68	1
5–10	35.67	2
10–20	20.82	3
20–30	14.01	4
> 30	2.83	5

Table 1.
Slope classification.

topographic instability conditions [7]. To characterize land use in this study, the Landsat 8 OLI satellite image sensor acquired on December 18, 2021 in our study area (WRS_PATH = 186 and WRS_ROW = 56) is the main data used. It was analyzed using Envi 5.3 image processing software, taking field observations into account. The classification of the different land use units was carried out depending on whether or not it promotes erosion (**Table 2**).

2.3.1.3 Soils

Soil data are a control variable for the erosion process. Its participation in the erosion phenomenon depends on its permeability and the ability to detach and transport its particles. Each soil type will react differently to the attack of rain and shear of runoff, depending on its texture, structure, porosity, and level of organic matter. The Menoua watershed is made up of a mosaic of soils linked to the geological history of the region. They come from basalts, sandy soils, and alluviums. To assess erodibility, we took into account not only the infiltration capacity which, according to Ref. [21], allows us to know the soil runoff potential, but also texture and organic matter content which, according to Ref. [22], have a considerable influence on the sensitivity of soils to erosion. These factors condition the permeability and cohesion of aggregates. According to these criteria, four major classes of soil are thus defined for the Menoua basin. Ferralitic soils, which benefit from good internal drainage form the first class and have a high infiltration capacity. In the second class, we have soils rich in humus and poorly developed soils. They generally have lower percolation and infiltration rates. The third class consists of moderately organic soils, such as poorly drained fine sands, loamy soils, and thin permeable soils. Hydromorphic soils consist of poorly structured and poorly drained heavy clays, which are found in the fourth class. According to these criteria, two major soil classes are defined for the Menoua basin. The first class is made up of ferralitic soils with good internal drainage and very high infiltration capacity. The second class includes hydromorphic soils consisting of moderately organic, poorly drained, and permeable thin soils (**Table 3**).

2.3.1.4 Drainage density

Drainage density is an input parameter into the various soil water erosion models, which represent tools to help implement future soil conservation plans. It is indicative of the infiltration and permeability of the basin. A high drainage density reflects the

Land use	Surface (%)	Influence on erosion
Agricultural zone	47.12	4
Built up area	4.81	3
Herbaceous savannah	32.21	2
Shrubs	15.86	1

Table 2.
Landuse distribution.

Soil type	Class	Surface (%)	Influence on erosion
Humic ferralitic soil on basalt	Ferralitic soil	6.53	5
Typical indurated ferralitic soil on basalt		1.93	4
Ferralitic soil is typically red on basalt		52.08	3
Typical ferralitic soil on gneiss		32.72	2
Moderately organic hydromorphic soil	Hydromorphic soil	6.73	1

Table 3.
Soil classification according to their contribution to erosion.

impermeable lithological nature that favors surface runoff. The hydrographic network of the Menoua basin is of order 6; therefore, very active during the rainfall.

2.3.1.5 Climate

Climate is an important factor that directly or indirectly influences soil erosion [23]. Rainfall in the humid tropics is the most important climatic variable that affects soil erosion. The action of rainfall amplifies the driving forces necessary for the uprooting of soil particles. Rainfall intensity and energy trigger soil erosion. The precipitation map of the Menoua watershed was produced using rainfall data from two meteorological stations (IRAD-Dschang, Djutitsa). These data have undergone interpolation operations (spline interpolation).

The sensitivity factors resulting from these data were rasterized in dimension 10 m*10 m for a pixel in order to harmonize the spatial resolution. These rasterization operations were carried out using the ESRI software range (ArcGIS© 10.3) at the URCLIEN Research Unit at the Department of Geography of the University of Dschang.

2.3.2 Hierarchical structuring of factors

Establishing the hierarchical structure consists of classifying the various factors selected according to their degree of influence on soil erosion. To facilitate the task, [16] set up a scale of numerical values (**Table 4**).

Values	Numerical scale for a comparative judgment of the indicators
1	Of equal importance
3	A little more important
5	Most important
7	Truly more important
9	Absolutely important
2,4,6 et 8	Values associated with intermediate judgments

Table 4.
Scale proposed by Ref. [16].

Factors	Numerical scale for a comparative judgment of indicators																	Factors
Slope	9	8	7	6	5	4	3	2	1	2	3	4	5	6	7	8	9	Land use
Slope	9	8	7	6	5	4	3	2	1	2	3	4	5	6	7	8	9	Land use
Slope	9	8	7	6	5	4	3	2	1	2	3	4	5	6	7	8	9	D density
Slope	9	8	7	6	5	4	3	2	1	2	3	4	5	6	7	8	9	Rainfall
Land use	9	8	7	6	5	4	3	2	1	2	3	4	5	6	7	8	9	Soil Type
Land use	9	8	7	6	5	4	3	2	1	2	3	4	5	6	7	8	9	D density
Land use	9	8	7	6	5	4	3	2	1	2	3	4	5	6	7	8	9	Rainfall
Soil type	9	8	7	6	5	4	3	2	1	2	3	4	5	6	7	8	9	D density
Soil type	9	8	7	6	5	4	3	2	1	2	3	4	5	6	7	8	9	Rainfall
D density	9	8	7	6	5	4	3	2	1	2	3	4	5	6	7	8	9	rainfall

NB: D density = Drainage density.

Table 5.
Comparison of factors by the expert.

2.3.3 Elaboration of binary combinations

After taking the advice of some researchers and experts on the study of water erosion of soils, the binary combinations, which consist of comparing the factors of erosion with each other within a matrix and assigning to each pair a comparison coefficient were made (**Table 5**).

The values in red are those checked by the expert to materialize the existing links between the factors to be compared. For example, considering the fourth line, the "Slope" indicator is really more important than the "rainfall" factor in the evaluation of the sensitivity of soils of the Menoua watershed to erosion. From this comparison between the different factors, a reciprocal comparison matrix was produced (**Table 7**) by applying the following relationship Eq. (1):

$$A = [aij] with \begin{cases} aii = 1 fori = 1..Ket \\ aji = \frac{1}{aij}(reciprocalvalue) \end{cases} \quad (1)$$

2.3.4 Determination of the value and the proper vector for each indicator

The weighting of the criteria makes it possible to reflect the relative importance given to each criterion by the experts. Once the comparison matrix has been obtained, the proper value of each combination and the proper vector corresponding to it are determined. The proper value of each pair comparison is obtained by dividing the numerical importance assigned to the pair by the sum of the numerical degrees of importance of the column. The proper vector indicates the order of priority or the hierarchy of the vulnerability indicators studied. It indicates the relative importance of the indicators. It is estimated by first calculating the sum of the proper values contained in each row of the matrix, then dividing this value by the number of indicators contained in the matrix. The proper vector associated with each factor is the weight assigned to each factor. Calculating the weight of each factor normalizes the comparison matrix so that the sum of all the weights equals 1.

2.3.5 Study of coherence of judgment

Computed priorities make sense only if the matrix of comparison by pairs is coherent. The evaluation of the coherence of judgments can be made using a Coherence Index (CI). This index measures the logical coherence of judgments of the people consulted. It provides information on consistency in terms of the ordinal importance of indicators to be compared. The estimation of this index is based on the calculation of the proper values of the comparison matrix using the mathematical procedure Eq. (2);

$$IC = (\lambda_{max} - n)/(n - 1) \tag{2}$$

such that λ_{max} is the maximum proper value of the comparison matrix, obtained by multiplying the total of each column of the comparison matrix by pairs with the relative weight of the indicator of this column, and by adding the results obtained for each column. n is the number of indicators compared in the matrix. The consistency ratio (CR) is then calculated, such as Eq. (3);

$$CR = \frac{IC}{IA} \tag{3}$$

where IA is the random index fixed according to the number of factors (4 in the case of this study). The value of AI was given by Ref. [16], and it is a function of the number of elements compared (**Table 6**).

If CR is less than or equal to 0.1 or 10%, then it is accepted that the weights assigned to the indicators are acceptable and, therefore consistent. If this threshold is exceeded, we are in a situation of inconsistency, then the matrix resulting from the comparisons will have to be reevaluated.

2.3.6 Aggregation of weighted data

This final stage of the Hierarchical Multi-criteria Analysis (HPA) occurs once the weighting of the landslide assessment factors has been carried out. At this point, it is easy to combine them to obtain an assessment of the sensitivity of the watershed to erosion. The most common and well-known technique of this approach is the weighted linear combination, which integrates all the considered factors into one [24–26]. It consists of multiplying each layer factor by its respective weighting coefficient, and then adding these results to produce a sensitivity index. The mathematical transcription of this combination is expressed as follows Eq. (4);

$$V_i = \sum_{j=1}^{5} \omega_j . a_{ij} \tag{4}$$

Number of indicators	1	2	3	4	5	6	7	8	9	10	11
IA	0	0	.58	.90	1,12	1,24	1,32	1,41	1,45	1,49	1,51

Table 6.
Random consistency indices [16].

where V_i is the summary index of susceptibility, ω_j is the weight attributed to each indicator, and a_{ij} is the weighting coefficient evaluating the relative importance of the factors.

This methodology for modeling and mapping soil sensitivity to erosion, which takes into account not only the functioning of the entire system but also the interrelationships between its various factors, is shown diagrammatically in **Figure 2**.

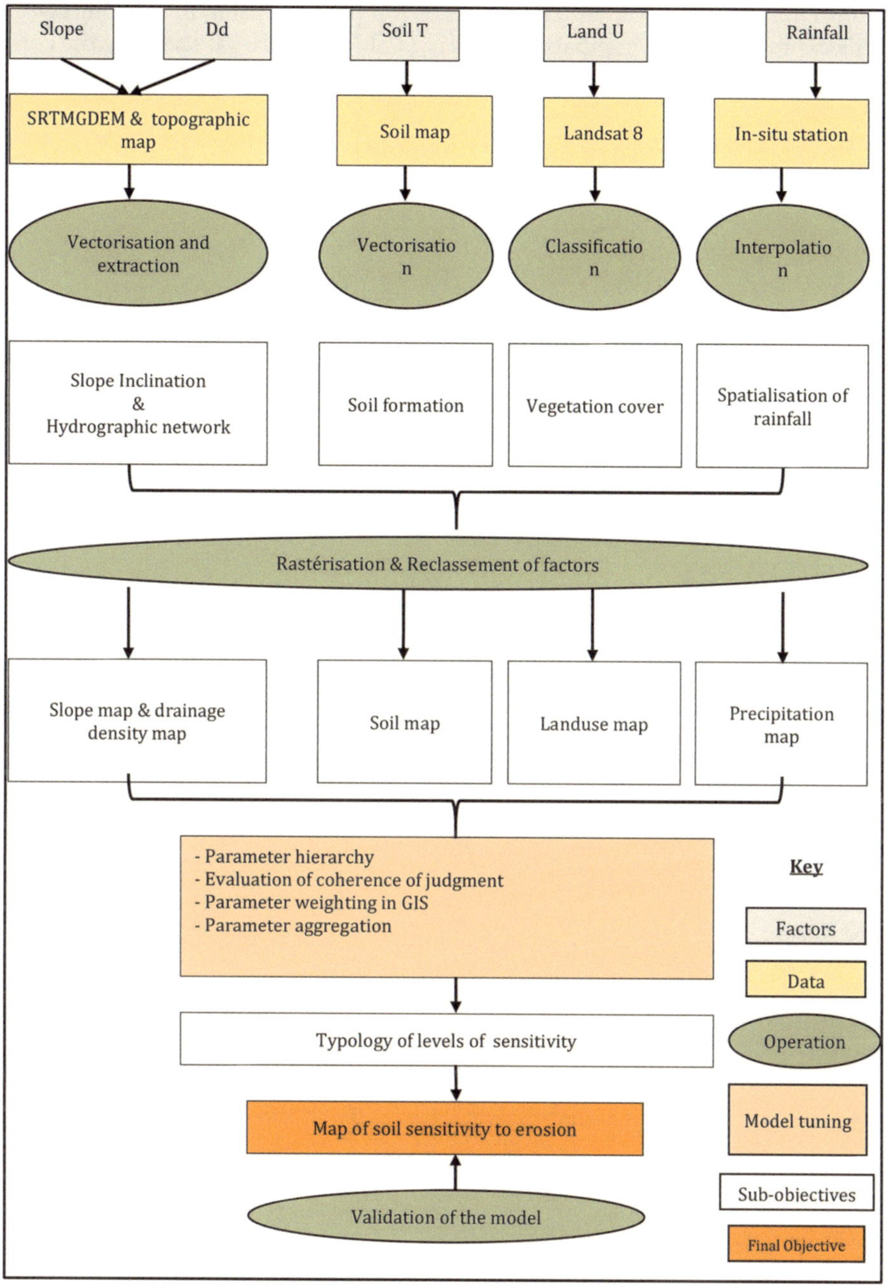

Figure 2.
Methodology for mapping soil sensitivity to erosion using the multi-criteria assessment.

3. Results

3.1 Analytical maps

The production of five analytical maps (**Figure 3**) was necessary to obtain the erosion sensitivity map.

Once the maps were produced, we used experts' assessment to compare in a reciprocal matrix, the different parameters of soil predisposition to erosion. The weights expressed by the comparison matrix (**Table 7**) were validated by the consistency of judgment in accordance with the equation .... and appended to the various maps produced.

According to the judgments made at the level of the comparison matrix, it is noted that the soil cover plays the most important role in the occurrence of erosion.

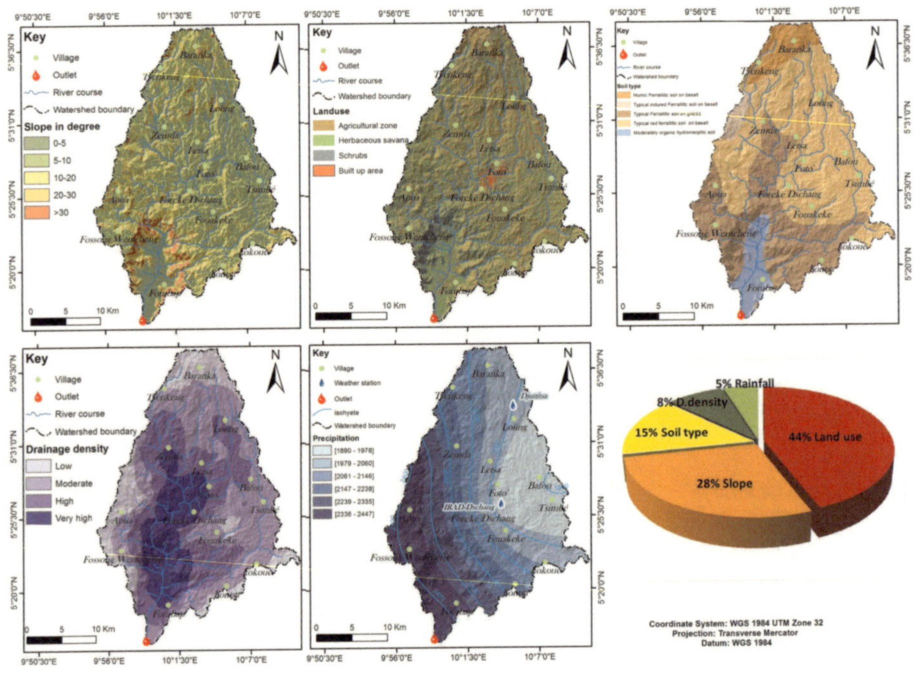

Figure 3.
Predisposition parameters of the cartographic model.

Factors	Slope	Land use	Soil Type	D.d	Rainfall	Weights (ω)	Weights (ω)%
Land use	1	2	4	5	7	0.44	44
Slope	1/2	1	3	4	5	0.28	28
Soil type	1/4	1/3	1	3	4	0.15	15
D.d	1/5	¼	1/3	1	2	0.08	8
Rainfall	1/7	1/5	1/4	1/2	1	0.05	5
				RC = 0,041			

Table 7.
Comparison of weights relative to factors of sensitivity.

According to experts consulted, the effectiveness of a downpour depends on the degree of protection that the plant cover can provide to preserve the soil. This led them to consider the role of the land use factor with a score of 44%. Secondly, the slope factor contributes about 28% to the occurrence of erosion. Indeed, runoff acts when the slope becomes steeper on soft soils.

However, on the steep slopes of Bamboutos Mountains, the technique of cultivation by plowing, which is the most widespread, helps to crumble the soil into fine particles that are easily transported by runoff water along steep slopes. For these experts, the structural stability that determines the soil's susceptibility to erosion depends on organic matter, aggregation, and texture. The result of the weighting obtained displays a score of 15% for the soil type factor. In general, the sensitivity to erosion in the Nkam watershed from the pedological point of view increases from moderately organic hydromorphic soils to humic ferralitic soils on basalt and shows on the whole that these soils are less resistant to erosion. The triggering factor precipitation (5%), which is the sinequanone condition for water erosion precedes that of drainage density (8%) that conditions diffuse erosion and generates the formation of capping crusts by reducing soil infiltration capacity.

3.2 Mapping of soil sensitivity to erosion

Once the final matrix was produced, the weights (ω) expressed by the matrix were associated with each thematic map in GIS using reclassification functions. Then, the cartographic algebra operations were carried out by applying Eq. (4). The usual formats for these calculations are rasters. As such, the resultant is a soil erosion sensitivity map for the Menoua watershed (**Figure 4**).

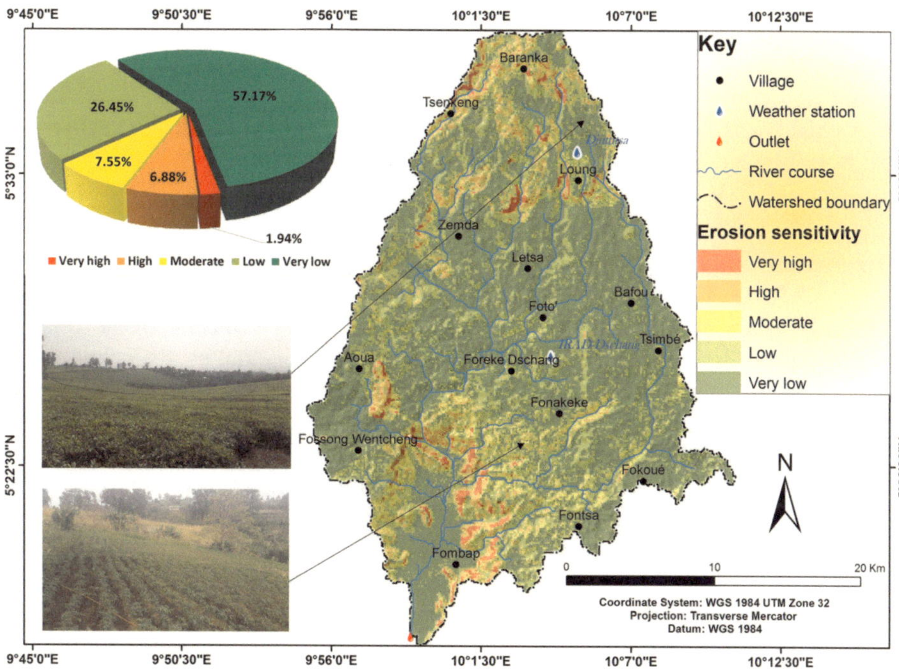

Figure 4.
Potential sensitivity of the Menoua watershed to erosion.

The analysis of **Figure 2** and the statistical distribution of the zoning of soil sensitivity to erosion in the Menoua watershed shows that levels of very strong and strong erosion sensitivity represent 8.82% or 5592.26 hectares of the total area. Field observations show that these classes of erosion are characterized by constantly plowed bare ferralitic soils on steep slopes, highly perceptible in the northern (toward the summit of the Bamboutos Mountains) and southern (on the Foréké escarpment which separates the Bamileke plateau from the Mbo plain) parts of the watershed. The moderate erosion sensitivity level covers 4781.31 hectares or 7.55% of the basin and is explained by the coincidence of medium slope classes with areas with a reduced vegetation cover, especially in the north, west, and southeastern parts of the watershed. Finally, the class of low and very low erosion status of 52981.91 hectares (i.e., 7% of the study area) occupies the center of the basin. Here, erosion is less advanced due to a fairly large vegetation cover and limited influence of the slope gradient.

3.3 Model validation

Any model must be validated so that the mapping of risks does not lead to risks for the mapping system [27]. Since no previous survey of the study area has relied on quantitative methods, this makes confirmation of results difficult. The model presented (**Figure 3**) was validated thanks to a reconnaissance mission of the mapped sites. Therefore, the validation of this model required field measurements and observations.

GPS surveys of certain bare slopes were carried out. It appears from this mission that all the sites mapped do not fully reflect the reality on the ground. For each test value, the two information plans (GPS surveys/model) were superimposed using GIS. Some sites identified as prone to erosion represent a rate of 73.87% coincidence with the model. Moreover, for about 26.13% of the results obtained, the model did not agree and sometimes very largely with field measurements (**Figure 5**). Sites classified as very sensitive are those on which we can either carry out reforestation schemes or no-till farming techniques.

4. Discussion

The assessment of soil sensitivity to water erosion using a multi-criteria approach calls on several criteria which, when put together, makes it possible to establish a balance sheet and a diagnosis of the physical degradation of soils in a given area. This study employed five criteria (slope, land use, lithology, drainage density, and precipitation) whose weighting was ensured by the AHP method. This method envisages a good understanding of the criteria involved in the erosion process, based on the pair-wise comparison of criteria at the same hierarchical level [28]. Land use was the most determining criterion in the process. Results obtained from the weighting of the criteria corroborate those of Refs. [29–31], in which plant cover protects the soil against ablation and reduces risk of erosion. Vegetation, therefore, acts as a protective screen against aggressive climatic conditions by intercepting the energy released by raindrops [32]. The topographic factor through the rigidity of the slope accentuates the erosive force of sheets' flowing water. But its behavior with respect to infiltration depends on the type of soils [33].

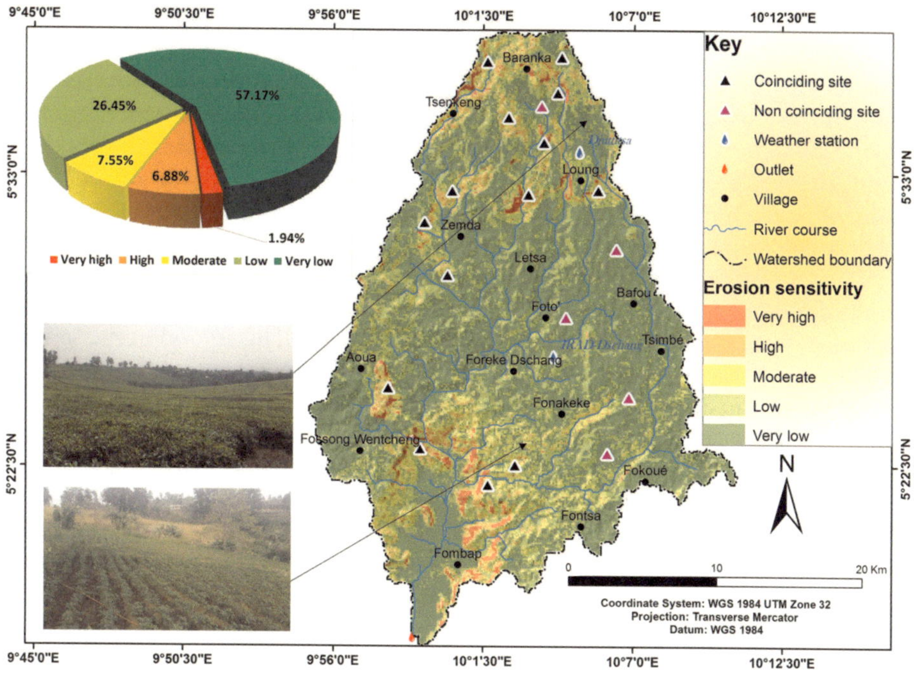

Figure 5.
Overlay of the different information planes (model & field data).

However, the model produced on the basis of these five criteria after validation gave satisfaction at a rate of 73.87%. The non-satisfaction of the model (23.13%) stems from the input data used in this study. They are mostly medium resolutions. One of the parameters whose imprecision impacted the final result was land use. Indeed, the satellite image used to establish the land cover map has a spatial resolution of 30 m x 30 m. This spatial resolution certainly did not allow more detailed observation of the earth's surface. The consequence of this imprecision on the result of this study is that the soils which could have been identified as less sensitive to water erosion were not, and *vice versa*. Forests with small dimensions may not have been taken into account in the analyses. The topographic data used to produce the slope and drainage density maps were generated with a scale of 1/50,000, which does not provide excellent precision. These inaccuracies in the model's input data can therefore influence the final result obtained in this study.

Moreover, the AHP method used for the weighting of the criteria, although relatively efficient, presents difficulties. One of the difficulties of this method concerns the subjectivity in the choice of the scale of scores ranging from 1 to 9 with their reciprocal correspondence [34]. The choice of the score corresponding to a criterion is arbitrary and may influence the calculation of the weight of the criterion considered. However, this method (AHP) is appropriate for this study because it performs well when the number of actions is reduced [35]. Nevertheless, despite these imprecisions on the input data, the quality of the model proves to be satisfactory for simulating the sensitivity of soils to erosion, which is necessary for any spatial planning study. It could be improved through the acquisition of high spatial resolution data.

5. Conclusion

Associating multi-criteria analysis with GIS offers possibilities for risk management that integrates, in a systemic vision, a set of parameters relating to its comprehension. These techniques have been applied to the Menoua watershed for modeling soil sensitivity to erosion. The final map obtained is intended to guide actors on the sustainable management of potential erodible zones. It highlights five categories of sites: very low, low, moderate, high, and very high sensitivity to erosion. There are two sites whose sensitivity to soil erosion is high. The first to the north, is around the summit of Mount Bamboutos, where a combination of ferralitic soils, agricultural activities, and steep slopes are decisive factors of erosion. The second in the south, is limited to the major tectonic zones of the area (foréké escarpment). It has a high drainage density in areas whose outcropping lithology corresponds to ferralitic soil formations on gneiss that stand out on the sloping ground. As for areas with low erosive sensitivity, they are found in isolated points throughout the basin. The study reveals that the level of very strong and strong erosion sensitivity accounts for 8.82%. The basin is therefore a geographical area at risk of erosion. In order to optimize the reliability of the map produced, a field mission made it possible to validate the sites prone to erosion. Based on these findings, it is recommended that erosion sensitivity should be taken into account when carrying out agricultural development projects. Adopting no-tillage farming technique and the agroforestry can reduce sensitivity to erosion and ensure sustainable management of mountains.

Acknowledgements

We express sincere thanks to late Professor TSALEFAC Maurice (Former-Dean of the Faculty of Letters and Human Sciences of the University of Dschang), who created favorable and essential conditions for the realization of this research work. He took upon himself the responsibility to supervise our academic works from the Masters to the Doctorate level. May his soul finds in this chapter, the expression of our deepest gratitude and infinite recognition.

Author details

Gabriel Nanfack* and Moye Eric Kongso
University of Dschang, Dschang, Cameroon

*Address all correspondence to: gabriel.nanfack@yahoo.fr

References

[1] Matsaguim NCA, Tiomo Dongfack E, Ngoufo R. Méthodesimplifiée pour la cartographie de l'occupation du sol en zone de montagne: cas du bassin versant de la Menoua (Région de l'Ouest-Cameroun). Revue Scientifique et Technique Forêt et Environnement du Bassin du Congo. 2019;**13**:21-33. DOI: 10.5281/zenodo.3518832

[2] Djoukeng HG. Le billonnage cloisonné en agriculture des montagnes: évaluation et facteursd'acceptation. Cas des HautsPlateaux de l'Ouest-Cameroun [thesis]. Université de Liège - Gembloux Agro-Bio Tech; 2016

[3] Derancourt F. Erosion par ruissellement des terresagricoles, méthodologieproposée à l'étude des bassins versants agricoles. Rapport Chambred'Agriculture Pas-de-Calais. 1995

[4] El Hage HH, Touchart L, Faour G. La sensibilitépotentielle du sol à l'érosionhydrique dans l'ouest de la Bekaa au Liban. M@ppemonde. 2013;**9**:1-7

[5] Kouedjou IL, AnabaBanimb RC. Analysemorphométrique du bassin versant de la Menoua pour unemeilleure gestion des risques morpho hydrologiques. American Journal of Innovation Research Application Science. 2021;**12**:188-197. Available from: http://creativecommons.org/licenses/by-nc/4.0/

[6] Kengni L, Tekoudjou H, Tematio P, Pamo TE, Tankou CM, Lucas Y, et al. Rainfall variability along the southern flank of the Bambouto Mountain (West-Cameroon). Journal of the Cameroon Academy of Sciences. 2009;**8**:45-52

[7] Nanfack G, Sone Diabe E. Évaluation de la susceptibilité des sols au ruissellement dans le bassin versant du Nkam (Ouest-Cameroun), en appliquant la méthode Service de Conservation des Sols Numero de Courbe. Romanian Association of Geomorphologists. 2021;**23**:39-53. DOI: 1021094/rg.2021.136

[8] Nanfack G, Julius Tata N. Mapping using multi-criteria hierarchical analysis of landslide risk areas in Menoua watershed (West-Cameroon). Geomatics-Planning-Development. 2020;**7**:56-71

[9] Tchoua FM. Contribution à l'étude géologique et pétrologique de quelques volcans de la ligne du Cameroun (Mont Manengouba et Mont Bamboutos) Cameroun. France: Université de Clermont Ferrant; 1974

[10] Aboubakar B, Kagou A, Nkouathio D, Ngapgue F. Instabilités de terrain dans les hautesterres de l'Ouest Cameroun : caractérisationgéologique et géotechnique du glissement de terrain de Kekem. Bulletin de l'Institut Scientifique, Rabat, Section. Sciences de la Terre. 2013;**35**:39-51

[11] Kaffo C, Fongang G. Les enjeuxagricoles et sociétaux de l'eau sur les montsBamboutos (Cameroun). Cahiers Agricultures. 2009;**18**:17-25

[12] Tchatchouang FC, Djamfa RC, YoutaHappi J, Tchawa P, Grozavu A. Expansion des cultures de contre-saison, changementsd'utilisation du sol et les implications environnementales dans les paysanneries de l'ouest-Cameroun, Scientific Annals Of "AlI. Cuza". University of Iaşi. 2014;**30**:41-57. DOI: 10.15551/scigeo.v60i2.316

[13] Jinghu P, Yan W. Estimation of soil erosion using RUSLE in Caijiamiao

watershed, China. Natural Hazards. 2014;**71**:2187-2205

[14] Kefi M, Yoshino K, Zayani K. IsodaH: Estimation of soil loss by using combination of Erosion model and GIS-case of study watersheds in Tunisia. Journal of Arid Land Studies. 2009;**19**: 287-290

[15] Sadiki A, Bouhlassa S, Auajjar J, Faleh A, Macaire J. Utilisation d'un SIG pour l'évaluation et la cartographie des risquesd'érosion par l'Equation universelle des pertesen sol dans le Rif oriental (Maroc):cas du bassin versant de l'ouedBoussouab. Bulletin de l'InstitutScientifique, Rabat, section Sciences de la Terre. 2004;**26**:69-79

[16] Saaty TL. A scaling method for priorities in hierarchical structures. Journal of Mathematical Psychology. 1977;**15**:234-281

[17] Jankwoski P. Integrating geographical information systems and multiple criteria decision-marking methods. International journal of geographical. Information Systems. 1995;**9**:251-273

[18] Hudson NW. Soil Conservation. London: Bastford; 1973. p. 320

[19] Roose E. Introduction à la GCES. Rome: Bulletinpédologique FAO; 1994. p. 420

[20] Wall GJ, Coote DR, Pringle EA, Shelton IJ. RUSLE-CAN Équation universelle révisée des pertes de sol pour application au Canada. Manuel pour l'évaluation des pertes de sol causées par l'érosionhydrique au Canada. Direction générale de la recherche, Agriculture et Agroalimentaire Canada, N° de la contribution AAC2244F. 2002;**2002**. p. 117

[21] Gassman PW, Reyes MR, Green CH, Arnold JG. The soil and water assessment tool: Historical development, application, and future research directions. American Society of Agricultural and Biological Engineers. 2007;**50**:1211-1250

[22] Wischmeier WH, Johnson CB, Cross BV. Un monogrammed'érosion du sol pour les terresagricoles et les chantiers de construction. Journal de la conservation des sols et de l'eau. 1971;**26**: 189-193

[23] Toy TJ, Foster GR, Renard KG. Erosion des sols: Processus, prediction, mesure et contrôle. New-York; 2002. p. 338

[24] Malczewski J. GIS-based land-use suitability analysis: A critical overview. Progress in Planning. 2004;**62**:3-65

[25] Conchita MGK. Modélisation géomatique par évaluation multicritère pour a prospection des sites d'agricultureurbaine à Ouagadougou. Revue électroniqueen sciences de l'environnement. 2010;**2010**:10. Available from: http://vertigo.revues.org/10368

[26] Feizizadeh B, Blaschke T. Land suitability analysis for Tabriz County, Iran: A multi-criteria evaluation approach using GIS. Journal of Environmental Planning and Management. 2012;**26**:1-23. DOI: 10.1080/09640568.2011.646964

[27] Cornelis B, Billen R. La cartographie des risques et les risquesde la cartographie1. documents. Tips. 2021;**2**: 207-222

[28] Jiang X, Lu WX, Zhao QG, Yang QC, Yang ZP. Évaluation des risques écologiques potentiels et prévision de la

pollution des sols par les métaux lourds autour de la décharge de la gangue de charbon. Natural Hazards Earth System Sciences. 2014;**14**:1599-1610. DOI: 10.5194/nhess-14-1599-2014

[29] Sadiki A, Faleh A, Zêzere JL, Mastass H. Quantification de l'érosionen nappes dans le bassin versant de l'ouedSahla Rif central Maroc. Cahiers Géographiques. 2009;**12**:59-70

[30] El Garouani A, Chen H, Lewis L, Tribak A, Abahrour M. Cartographie de l'utilisation du sol et de l'érosionnette à partird'imagessatellitaires et du SIG IDRISI au nord-est du Maroc. Revue télédétection. 2008;**8**:193-201

[31] el Hage H, Charbel L, Touchart L. Modélisation de l'érosionhydrique à l'échelle du bassin versant du Mhaydssé. Békaa-Liban. Vertig0- la revue électroniqueen sciences de l'environnement. 2018;**2018**:18. DOI: 10.4000/vertigo.19804

[32] Boukheir RG, Khawlie M, Abadallah C. Erosion hydrique des sols dans les milieuxméditerranéens. Revue bibliographique, érosion et gestion des sols. 2021;**8**:231-245

[33] Casenave A, Valentin C. Les états de surface du sol dans la zone sahélienne. Paris: ORSTOM; 1989. 229 p

[34] Belton V. A comparison of the analytic hierarchy process and a simple multiattribute value function. European Journal Operational Research. 1986;**26**: 228-230

[35] Laaribi A. SIG et analysemulticritère. Paris: Hermès Science Publications; 2000. 196 p

Chapter 5

Soil Erosion Risk Analysis of a Small Watershed

Charles Galdies, Amy Zammit and Adam Gauci

Abstract

Malta is being rapidly exposed to developmental activities occurring inland and along its coastline, which in turn triggers erosion and flooding in the event of high-intensity rainfall. Most of the rainwater-containing several contaminants from urban and agricultural areas are lost as runoff into the coastal waters, which in turn have adverse environmental and socioeconomic impacts. The extent of soil erosion and runoff can be investigated starting from the watershed basin downhill till coastal waters. This study links the runoff of soil along an ecologically sensitive watershed in Malta with the use of multidisciplinary techniques. These included the estimation of soil erosivity coupled with satellite remote sensing chlorophyll-a (CHLA) and total suspended matter (TSM) in coastal waters adjacent to the mouth of the valley. This represents a novel study for the Maltese islands because it provides a precise map of soil erosion hotspots in the Ramla watershed as high as 30 ton ha^{-1} yr^{-1}. Using three case studies of past torrential rain episodes, the sedimentation process resulted in a 120% and 133% increase in CHLA and TSM levels, respectively, against background levels. This information is vital for proper risk management of ecologically sensitive watershed basins.

Keywords: Malta, RUSLE, chlorophyll-a, total suspended matter, COPERNICUS

1. Introduction

Finely textured sediments originating from upland areas get deposited further down in lowland areas, forming one of the most fundamental earth processes. Sedimentation processes stretch across other disciplines, including soil and plant science, geomorphology, and coastal zone management. Ongoing research in the field of sedimentology is, therefore, very relevant because changes in global sediment transport are being used as a primary form of evidence for the Anthropocene.

Like sedimentation, soil erosion is a dynamic, multistage process involving soil detachment, breakdown, transport, and subsequent deposition that is forced by wind or rain, or through soil-intensive human activities such as farming [1]. Soil erosion can lead to serious loss of topsoil and organic matter, which can lead to reduced vegetation growth and to biodiversity in general [2]. The focus of this chapter is on the multidisciplinary understanding between soil erosion, transport, and its impact within a localized watershed situated in the Maltese islands, using a range of techniques that are highly suited for risk management.

Figure 1.
The Maltese islands show the study area located in Gozo. (Source: OSM & Malta GeoPortal – Planning Authority).

The Maltese islands are located within the central Mediterranean region (**Figure 1**). According to CMIP6 climate models, the projected future climate of the islands is expected to experience a higher frequency of climatic extremes, including prolonged drought conditions and heatwaves, as well as increased frequency of torrential rain that can cause intermittent but significant flooding. The latter may, thus, result in increased mass wasting and related soil erosion, which can negatively affect erosion-vulnerable watersheds that are already significantly impacted by unsustainable and intense anthropogenic influences, including agriculture and associated sediment displacement. Clays and silts from upper parts of watersheds, once suspended in the stream network, can be routed directly to the mouth of the valley, resulting in significant sediment loads in coastal waters, and thus affecting exposed and submerged coastal ecosystems.

Severe rainstorms can have significant short- to long-term impacts on coastal water quality as a result of soil erosion and sedimentation processes both during and after their occurrence, especially if they are well outside historically observed norms. Extreme rainfall events can transport significant amounts of suspended sediment containing a variety of pollutants (such as heavy metals and organic compounds) through watersheds, ending up into coastal waters. Water runoff may also carry sufficiently coarse sediments to block the penetration of light through the water column, and thus be able to impact sensitive benthic flora and fauna. According to Ref. [3], the upper tolerance of total suspended sediments for most aquatic species is 80–100 mg/L, but it can be much less for bottom-dwelling aquatic invertebrates.

For these reasons, it is very important to understand the degree of these processes for eventual management of such risks within important watersheds. Such a study requires an accurate identification of the connectivity between the degree of soil erodibility and the corresponding impact to coastal waters resulting from sedimentation at the local scale. In this chapter, we look at an integrated assessment of sediment dynamics typically triggered by a rainstorm occurring within an important watershed in the Maltese islands, namely the identification of problems having a significantly high degree of soil erodibility, and the estimation of both total suspended matter (TSM) and chlorophyll-a (CHLA). The morphometrics of the Ramla watershed was derived from a very high-resolution DEM derived using LiDAR, and evaluated in detail. The revised universal soil loss equation (RUSLE) model was used to assess the degree of soil erodibility, while Earth observation technology was used to estimate the impact of surface runoff on coastal water.

2. Case study: The Ramla watershed in Gozo

The Ramla watershed in Gozo is roughly a 6 km^2 catchment area extending from its watershed divide down to the Ramla beach (**Figure 1**). The socioeconomic and environmental importance of this sensitive watershed is based on intensive farming activities [4], the presence of perched aquifers along its slopes (MT015 & MT016. [5]), unique biodiversity [6], and the presence of a NATURA 2000 site [4, 7].

According to Ref. [8], the Ramla watershed has formed through the fragmentation and dislodgment of the upper coralline plateau along the edges of the headlands overlooking, which is being sustained by the erosion of the underlying blue clay strata. The detachment of rock mass of various sizes then continues to fragment and weather into small pieces. Its geomorphology is characterized by somewhat steep relief, with elevations ranging between 120 and 28 m above sea level and an average slope of 18.1%. Its climate features mild, humid winters and hot, dry summers an annual mean air temperature of 18.6°C, and a mean precipitation of 574 mm [9]. The rainfall regime is intermittent with baseflow from October to January.

The Ramla watershed has been much affected by human activities. Agriculture has evidently shaped the watershed into a system of terraced fields that are delineated by rubble walls to define and protect the land parcels (**Figure 2b**). However, the area still continues to show natural depositional features including gullies and rills, as well as its principal watercourse and associated tributaries. The slopes are affected directly by rainfall resulting in surface runoff in the form of stream flow, which at higher slope elevations of the Ramla catchment removes the weathered material of the clay and makes the surface of the slope smoother. Splash erosion can happen when clay particles are moved about by raindrops. This becomes significant over the barren parts of slopes.

2.1 Methodology

2.1.1 Morphometric analysis

National 1 × 1 m DEM derived from LiDAR and 15 cm pixel resolution aerial orthophotographs [10] were used for morphometric analysis in a GIS environment. The terrain analysis approach of SAGA GIS V8.5.1 [11] was used to derive morphometry characteristics of the watershed. A number of specific software tools were used: channels [12], hydrology, and morphometry [12]. These tools extracted the presence of watersheds and sub-watersheds in an automated manner in the form of GIS vector layers. For a proper determination of flow direction and flow accumulation, sinks were identified from the DEM and filled. After the drainage networks were extracted and sub-watersheds were delineated, four sets of morphometric parameters were calculated using the mathematical formulations as described in **Table 1**.

The stream order is the primary step in quantitative interpretations of a drainage network, and the stream number is the number of streams in each order. These two

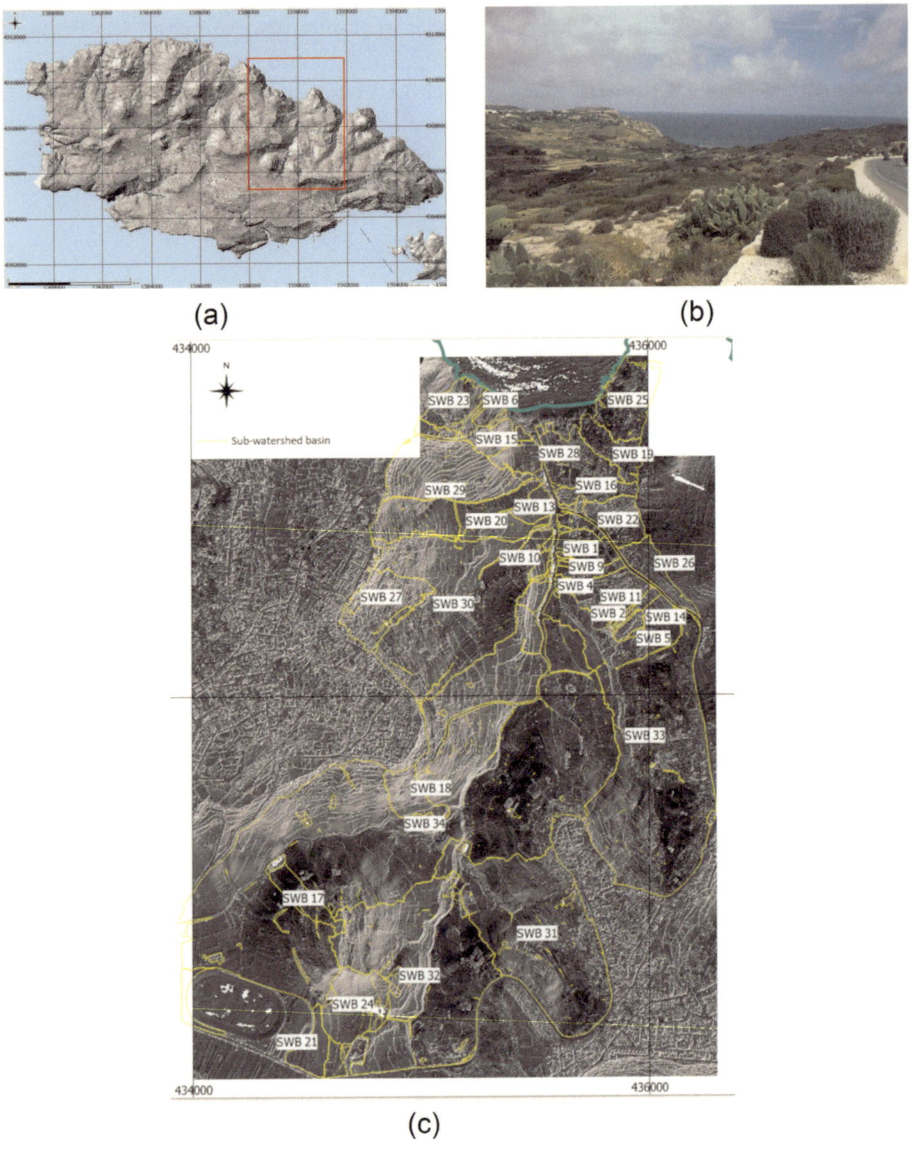

Figure 2.
(a) Topographic relief of the island of Gozo showing the area of interest (inset), (b) onsite agricultural land cover of the watershed taken from south-eastern watershed divide, and (c) analytical Hillshading derived from LiDAR-derived elevation data. (Source: Authors).

No	Parameters	Equation	References
1	Stream order	Hierarchic order	[13, 14]
2	Stream number	$Nu = N1 + N2 + N3 + \dots Nn = \Sigma N_u$	[13]
3	Stream length	$Lu = L1 + L2 + L3 + \dots Ln = \Sigma L_u$	[13, 14]
4	Mean stream length	$L_{ms} = Lu/Nu$	[14]
5	Terrain ruggedness index	$TRI=Y[\Sigma(x_{ij}-x_{00})^2]^{1/2}$	[15]

Table 1.
Parameters derived from the morphometric analysis.

parameters produce a geometric relationship that resulted in the calculation of the stream length of the drainage that was based on the surface runoff characteristics of the basin. The total length of individual stream segments of each order was, thus, derived, while the extracted Mean stream length provides information of the drainage size and its contributing basin surface. The terrain ruggedness index (TRI) was also calculated by calculating the sum change in elevation between a grid cell and its eight neighbor grid cells, where x_{ij} is the elevation of each neighbor cell to cell. The watershed basin characteristics, the channel network, and sink route were subsequently derived from the DEM to produce a basin shapefile, along with other shapefiles [12].

2.1.2 Soil erosion

Soil erodibility within the Ramla watershed was estimated using the RUSLE [16]. The soil erodibility was estimated using the following equation:

$$A = R^* \ K^* \ LS^* \ C^* \ P.$$

where A is the estimated soil loss in tons per hectare per year, R is the rainfall erosivity factor, K factor is a measure of soil erodibility, LS is the slope length and steepness factor, C is the cover and management factor, and P is the support practice factor.

2.1.3 Estimation of CHLA and TSM

The derivation of the climatological profile of CHLA and TSM was based on the analysis of 10 years' worth of monthly 300 m MERIS L1B sensor data derived by ENVISAT-1. The observation period ranged between May 17, 2002 until April 8, 2012, and is equivalent to 3614 days for the area of interest (**Figure 3**). The algorithm used to derive CHLA (in $mg.m^{-3}$), and TSM (in $g.m^{-3}$) was based on Case 2 water monitoring [17] with a spatial resolution of 0.3 km/pixel data products. Average concentrations of CHLA were derived using the OC4 algorithm and TSM [18]. The monthly climatological processing of the CHLA and TSM was conducted using MATLAB coding.

For this case study, near real-time, high-resolution COPERNICUS Sentinel-2A and 2B (L1C) imagery overpasses were used to derive CHLA and TSM quantitative maps that coincided with the rainstorm events observed between the January 2 and 12, 2021. Prior to processing, images were resampled using Band-2 due to its high spatial resolution (10 m). Salinity and temperature parameters needed for algorithm processing were set to reflect the values observed in Maltese coastal waters based on their monthly averages.

3. Results

3.1 Ramla watershed characteristics

This watershed shows highly interesting features as highlighted by the geomorphometric processing results that exploited the unprecedented spatial resolution provided by the national LiDAR data. **Table 2** shows the morphometric features of all the main hydrological sub-catchments that were derived using GIS-based terrain raster processing (**Figure 2c**). These features include the minimum, maximum, and

	Strahler stream number					Total stream length (m)					Mean stream length (m)				
Sub-Watershed basin	Order 1	Order 2	Order 3	Order 4	Order 5	Order 1	Order 2	Order 3	Order 4	Order 5	Order 1	Order 2	Order 3	Order 4	Order 5
1	1	0	0	0	0	54.4	—	—	—	—	54.4	0.0	0.0	0.0	0.0
2	2	2	0	0	0	87.5	82.9	—	—	—	43.7	41.5	—	—	—
3	0	0	0	0	0	—	—	—	—	—	0.0	0.0	0.0	0.0	0.0
4	0	0	1	0	0	—	—	1.0	—	—	—	—	64.8	—	—
5	2	1	0	0	0	31.0	30.7	—	—	—	15.5	30.7	—	—	—
6	2	0	0	0	0	174.7	—	—	—	—	87.4	—	—	—	—
8	0	0	0	0	0	—	—	—	—	—	0.0	0.0	0.0	0.0	0.0
9	2	1	2	0	0	144.4	25.1	41.3	—	—	72.2	25.1	20.6	—	—
10	1	1	1	0	0	23.0	29.1	39.0	—	—	23.0	29.1	39.0	—	—
12	5	2	2	0	0	172.1	102.6	38.0	—	—	34.4	51.3	19.0	—	—
13	0	0	0	0	0	—	—	—	—	—	0.0	0.0	0.0	0.0	0.0
14	13	5	0	0	0	430.8	296.0	—	—	—	33.1	59.2	—	—	—
15	10	4	6	0	0	494.4	220.5	128.4	—	—	49.4	55.1	21.4	—	—
16	9	6	1	0	0	219.3	296.5	12.1	—	—	24.4	49.4	12.1	—	—
17	13	3	8	0	0	477.1	92.2	540.8	—	—	36.7	30.7	67.6	—	—
18	20	17	0	0	0	646.6	966.0	—	—	—	32.3	56.8	—	—	—
19	19	13	1	0	0	1221.4	482.2	28.0	—	—	64.3	37.1	28.0	—	—
20	18	11	6	3	0	878.7	385.7	264.3	191.7	—	48.8	35.1	44.0	63.9	—
21	28	14	4	6	0	1446.3	815.0	147.9	262.0	—	51.7	58.2	37.0	43.7	—
22	26	14	6	0	0	1157.1	785.5	406.5	—	—	44.5	56.1	67.7	—	—
23	24	13	6	1	0	1039.4	521.7	144.3	165.8	0.0	43.3	40.1	24.1	165.8	—

Sub-Watershed basin	Strahler stream number					Total stream length (m)					Mean stream length (m)				
	Order 1	Order 2	Order 3	Order 4	Order 5	Order 1	Order 2	Order 3	Order 4	Order 5	Order 1	Order 2	Order 3	Order 4	Order 5
24	37	20	11	8	1	1434.5	968.2	327.9	319.9	54.7	38.8	48.4	29.8	40.0	54.7
25	42	22	6	0	0	1957.4	986.4	310.1	—	—	46.6	44.8	51.7	—	—
26	43	26	8	1	0	2162.4	678.8	162.7	30.6	—	50.3	26.1	20.3	30.6	—
27	54	36	12	0	0	1737.0	1511.4	403.8	—	—	32.2	42.0	33.6	—	—
28	39	26	7	0	0	2306.6	1892.7	374.9	—	—	59.1	72.8	53.6	—	—
29	50	23	19	0	9	1727.8	892.6	700.2	—	470.5	34.6	38.8	36.9	—	52.3
30	65	36	23	4	0	2132.0	1364.0	864.5	94.2	—	32.8	37.9	37.6	23.5	—
31	150	68	23	27	10	7514.4	2938.0	647.7	985.1	134.8	50.1	43.2	28.2	36.5	13.5
32	245	130	51	38	17	11427.3	5370.5	2209.2	1523.4	432.1	46.6	41.3	43.3	40.1	25.4
33	241	131	49	23	45	10820.9	5925.0	1874.3	904.2	818.6	44.9	45.2	38.3	39.3	18.2
34	283	135	68	36	19	11767.1	4899.2	2542.7	982.8	353.7	41.6	36.3	37.4	27.3	18.6

Table 2.
Calculated stream order, stream number, and stream lengths of the hydrological sub-watershed basins identified within the Ramla watershed.

Figure 3.
SENTINEL2-A image of the Ramla valley mouth, sandy beach, and adjacent coastal waters (Source: COPERNICUS).

average elevation of each sub-catchment basin together with the Strahler stream number and their lengths [13].

Detailed morphometric analysis shows that the largest sub-watershed basin is around 110 ha (SWB 3, **Figure 2c**). The mapping of the 4th and 5th stream order categories found in the Ramla watershed area is shown in **Figure 2b**. According to **Table 2**, the longest water channels belong to stream order 2 followed by stream order 1, with a total of 2760 and 1427, respectively. The drainage pattern is tightly linked to a landform, where rocks are flat-lying and preferential zones of structural weakness are minimal.

This integrated analysis points to a sub-watershed structure that is the direct result of an underlying lithology with a hydrology that is a direct function of the geomorphology and topography, including that which is artificially induced. Results show that stream runoff in the Ramla valley catchment is mainly dominated by SWB 34, 31, 33, and 10. Therefore, these sub-watersheds are deemed to be important from a soil erodibility management point of view.

3.2 Degree of soil erodibility from the Ramla watershed

Soil erodibility within the valley (**Figure 4b**) varies from a low (blue color shading) to a high rate of more than 30 ton ha^{-1} $yr.^{-1}$ (yellow color shading). The RUSLE estimation showed that the area around the in-Nuffara Hill is likely to be most affected by high erosion rates. This is because its slopes consist of fallow fields which lack proper maintenance of rubble walls together with the occurrence of motorcycle off-roading practice, which together are contributing to higher soil erosion. The high slopes along Xaghra (west) and Nadur (east) headlands, which consist of blue clay and also silty soil, also show a high degree of soil erodibility (**Figure 4b**).

The RUSLE estimation of the Ramla watershed shows that its soil erosivity can be generally considered as tolerable (<10 ton ha^{-1} $yr.^{-1}$). The type of land-use land

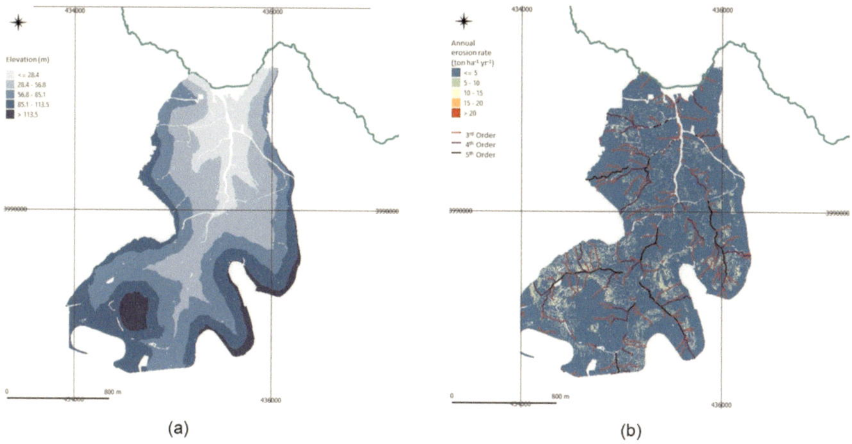

Figure 4.
DEM (a), and RUSLE, showing 3rd-, 4th-, and 5th-order channels derived by this study, leading to the mouth of the valley and into the coastal waters (b).

cover present within the tolerable range is generally characterized by seasonal crops and by a gentle slope gradient that tends to cause reduced runoff velocity. However, at higher slope gradients, one can observe elevated risks of soil erosion (>30 ton ha^{-1} $yr.^{-1}$) associated with the onsite presence of abandoned and/or neglected field terraces having poor vegetation cover. Site investigations also point to a significant collapse of protective barriers within these areas (**Figure 5**).

The terrain ruggedness index (TRI), which is based on the sum change in elevation between a grid cell and its neighboring grid cells of 1 m^2 ground resolution, reveals the distribution of the terrain heterogeneity. The resultant TRI map shows high ruggedness within certain parts of the valley system that are also prone to soil erosivity (**Figure 6b**). The TRI map shows high values to the left of the valley mouth, which incidentally is also associated with a significant degree of erodibility (**Figure 4b**, with areas having a soil erodibility of around 10–15 ton ha^{-1} $yr.^{-1}$) incorporating in it a number of stream channels that head directly toward the coast and therefore, contributing to sediment movement toward this direction. High index values are also

Figure 5.
*High risk of erosion located along the in-Nuffara Hill area (shown in red), which is primarily associated with SWB 34 (**Figure 3c**).*

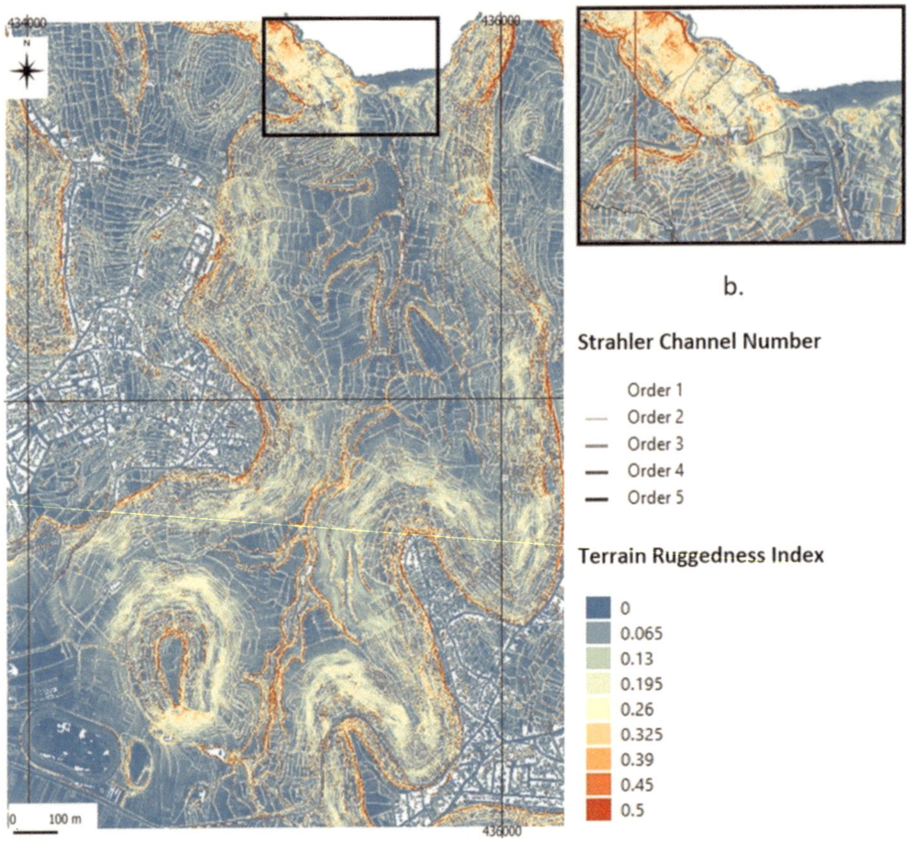

Figure 6.
*Terrain ruggedness index map derived for the Ramla watershed (a). Note the high index value in the inset (b) to the left of the valley mouth associated with significant degree of erodibility (**Figure 4b**, with areas having around 10–15 ton ha^{-1} yr.$^{-1}$) and associated stream channels heading toward the coast.*

found within the southern parts of the Ramla valley associated with the high degree of erosivity surrounding the in-Nuffara Hill (**Figure 5**).

3.3 Risk of sedimentation from the watershed into Ramla coastal waters following soil erosion due to rainstorms

3.3.1 Rainstorm events during January 2021

The daily rainfall amount in mm was obtained from the nearest weather station situated in Gozo (**Table 3**). Winds from the south reached an average speed 25 km/hr.

The computation and mapping of CHLA and TSM are shown in **Figures 7–13**. Histograms for coastal water quality showed that for the first part and end part of the event the CHLA is positively skewed (**Figure 8**), implying that concentrations returned to pre-rainfall conditions. The combined histogram (**Figure 9**) expressed the variation of concentrations before and after the rainstorm, which coincides with the climatological extent obtained for January based on MERIS data (**Figure 10**).

The TSM concentration showed a slight increase toward higher levels on the second date (**Figure 11**). However, the shift was less evident than for CHLA as frequency values remained positively skewed throughout the analysis (**Figure 12**). A slight increase in TSM can be seen for January 7th, implying that TSM concentrations had increased, likely due to sediment runoff from Ramla valley system (**Figure 13**).

Climatological analysis for the Ramla coastal waters shows that rainstorm events are highly likely to increase the water levels of CHLA and TSM levels since these two parameters were observed to peak during the wetter months. The near real-time high-resolution imagery confirmed this observation (**Table 4**) in a way that CHLA and TSM levels have increased after the rainfall event and decreased after a few days. This also links to the impact of rainfall on the sedimentation process within the Ramla watershed. According to **Table 4**, during the January rainstorm event, the sedimentation and associated nutrients ending up in adjacent coastal waters found next to the mouth of the watershed resulted in a 120% and 133% increase in CHLA and TSM levels, respectively, against background levels.

Date of rainstorm event	Rainfall (mm)
02/01/2021	8.5
03/01/2021	4.3
04/01/2021	3.2
07/01/2021	0.0
08/01/2021	0.3
12/01/2021	2.3

Table 3.
Rainfall (mm) during the rainstorm of January 2021.

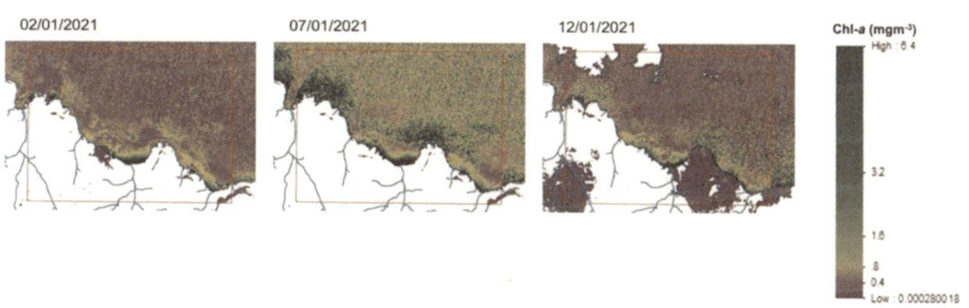

Figure 7.
CHLA concentration ($mg.m^{-3}$) during the period January 2–12, 2021 detected by SENTINEL2-A and B data. The rectangular outline shown in red defines the area of analysis.

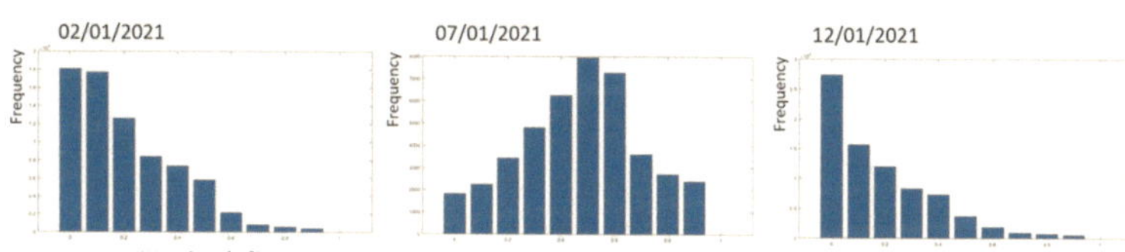

Figure 8.
Histograms for CHLA concentrations ($mg.m^{-3}$) during the period January 2–12, 2021 detected by SENTINEL2-A and B data.

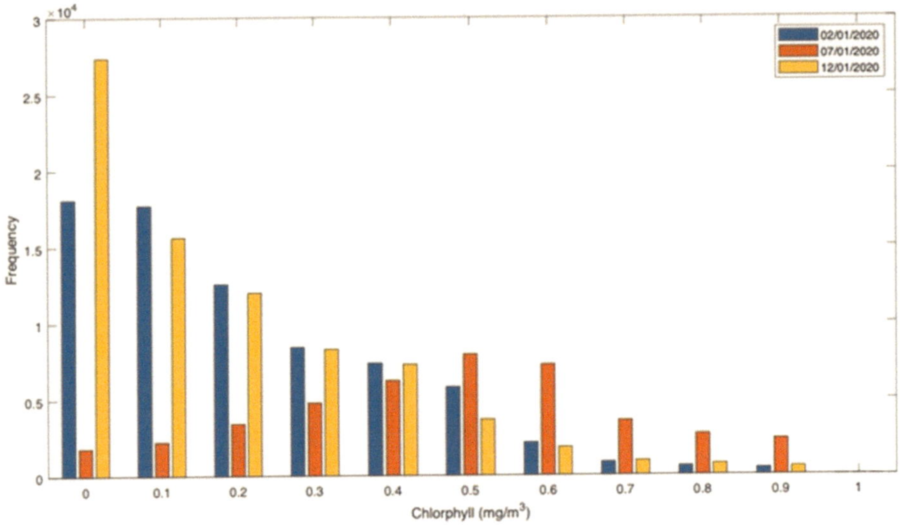

Figure 9.
Combined histogram of CHLA concentrations (mg.m^{-3}) for January 02–12, 2021 detected by SENTINEL2-A and B data.

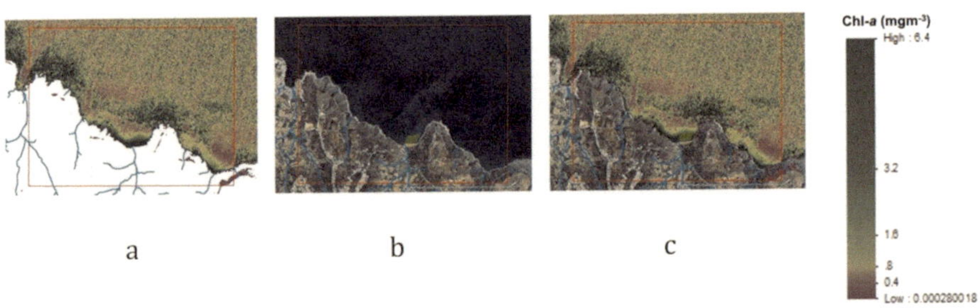

Figure 10.
SENTINEL2-A & B derived CHLA concentration (mg.m^{-3}) on January 07, 2021 (a); CHLA January climatology derived from MERIS data [14] (b); merged CHLA levels for the period January 02–12, 2021 (c). The rectangular outline shown in red defines the area of analysis.

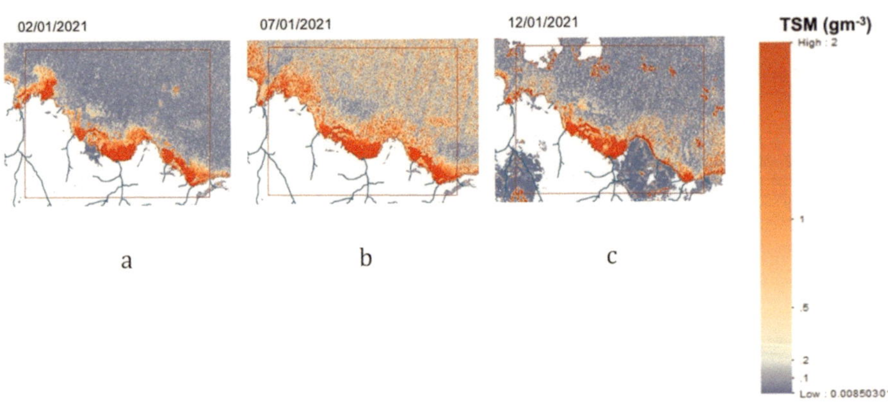

Figure 11.
SENTINEL2-A & B derived TSM concentration (g.m^{-3}) on 07/01/2021 (a); TSM January climatology (b); and merged TSM levels for the period January 02–12, 2021 (c).

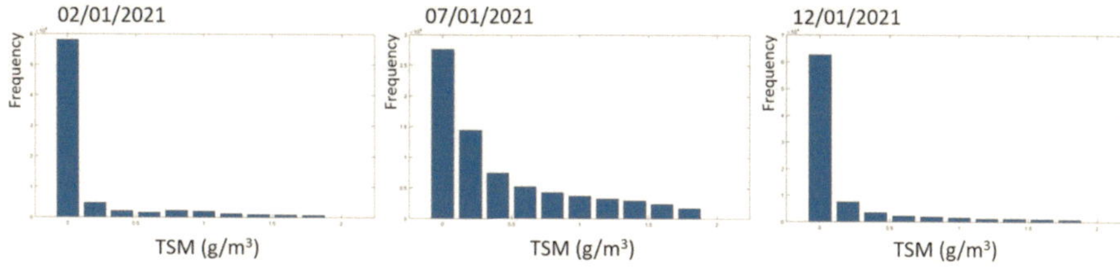

Figure 12.
Histograms showing TSM (g.m^{-3}).

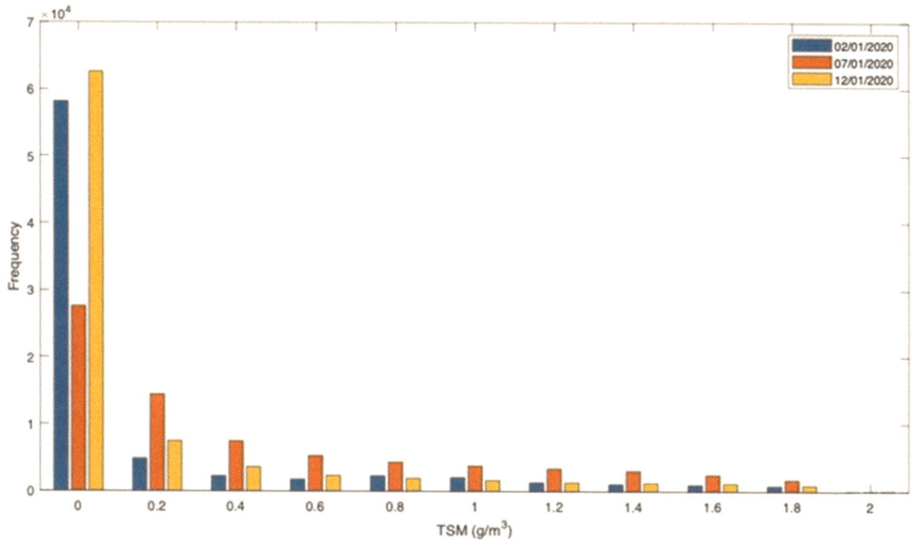

Figure 13.
Combined histogram of TSM (g.m^{-3}).

Date	Rainstorm event		Climatological levels for January	
	TSM (g.m^{-3})	CHLA (mg.g^{-3})	TSM (g.m^{-3})	CHLA (mg.g^{-3})
January 2, 2021	0.163	0.069	0.160	0.105
January 7, 2021	0.244	0.125	—	—
January 12, 2021	0.212	0.052	—	—

Table 4.
Comparative extent between the rainstorm events and the respective climatological levels for the Ramla coastal waters next to the valley mouth.

4. Discussion

This study presents a practical and multidisciplinary approach that is essential for risk management of watersheds because it helps identify the areas and factors that are most vulnerable to erosion and coastal sedimentation. Similar problems remain practically unexplored in Malta.

Soil erosion is a significant environmental issue for Malta and is having serious consequences for the sustainability of local watersheds and the agriculture sector that depends on them. Among other things, it leads to increased sedimentation along the drainage pattern identified by this study, and ultimately to increased risks of flooding and landslides within the watershed basin itself. In addition, unless properly managed, changes in the watershed dynamics will negatively affect the sensitive ecosystem dynamics observed at the Ramla valley mouth that are critical to its normal functioning.

From a risk management plan of this highly sensitive ecological watershed, it is being recommended that the impact of soil erosion hotspots identified by this study (both within the watershed itself and resulting impact on coastal waters next to the valley mouth), are to be further monitored, and if possible, analyzed, and mitigated. Sustainable management of watersheds necessitates the use of appropriate hydrological, ecological, and socioeconomic management tools that unfortunately tend to operate, if at all, in mutual isolation. Such a thorough and holistic approach would support the much-needed management of local watersheds, which is particularly concerning and urgent in view of the need for Malta to adhere to a number of important European directives, such as the Water Framework and the Marine Strategy Framework Directive.

Future studies related to coastal water sedimentation can consider using other Earth observations platforms, such as OLCI and LANDSAT 9, to analyze more rainstorms of varying severities. Ideally, such studies would include improved algorithms to detect CHLA and TSM in coastal waters associated with watersheds. Such an integrated monitoring approach, involving both *in situ* and remote sensing analysis must also be able to give value to the validation of the entire soil erosion and sedimentation process at a highly localized scale.

It is hoped that the analysis and quantification of this phenomenon will contribute to an understanding of the dynamics of an important watershed that so far has never been studied from a soil erosion risk management point of view. The results could be used in future multi- and inter-disciplinary studies to better understand the impact of unsustainable land use practices both within watersheds and related littoral sediment budgets and to manage and develop future risk management plans aimed at minimizing soil erosion under a changing climate over the Maltese islands where evapotranspiration, rainfall, and temperatures regimes will change significantly over the next 50 years [19].

The study has a number of methodological and logistical limitations that stem from its multidisciplinary nature. In the case of the estimation of the degree of soil erosion, scientific literature on this topic is filled with the application of RUSLE for relatively large areas and therefore, the present estimates should be interpreted with care. The biggest challenge, however, concerns the *in situ* validation of the degree of soil erosion mapping estimations as identified by the 3D GIS visualization. Unfortunately, no data on actual soil losses exist at the local scale against which the present estimates can be compared.

The sedimentation processes identified in the coastal waters just outside the valley mouth can be affected by inaccuracies due to differences in resolutions used between the near-real-time Sentinel-2 datasets (having a ground resolution of 10 m) against the MERIS imagery (with a ground resolution of 300 m) used to derive climatological baselines. The use of Sentinel data at the finest resolution constituted the best freely-available data available that can be used for our soil erosion and sedimentation study at the Ramla valley mouth.

The processing of Earth observation data also had its own limitations. Firstly, restricted satellite visiting times have led to limited availability of data over the Ramla watershed, especially when studying limited, high-intensity, low-duration rainfall events. A further limitation on the methodology was posed by the presence of clouds in the satellite scenes, which tend to contaminate the sensor information acquired over any area of interest. Naturally, clouds are present during rainfall and so optical remote sensing of coastal waters can be quite challenging during, and exactly following rainstorms.

5. Conclusion

Studying soil erosion is essential for appropriate risk management of small watersheds because it helps us to identify the areas and factors that are most prone to erosion. This study describes the identification of precise locations of soil erosion hotspots that fall within the high soil erosion category (30 ton ha^{-1} $yr.^{-1}$) within the small Ramla watershed. The impact of such an intolerable degree of soil erosion in this area is being demonstrated by following the resultant deterioration of coastal water quality at the mouth of the watershed during high rainfall episodes. Based on a case study, the impact of sedimentation resulted in a 120% and 133% increase in CHLA and TSM levels respectively against background levels. The observed CHLA levels during and after rainstorms indicate eutrophic water type resulting from this sedimentation. This study provides for the first time an understanding of the causes and effects of localized soil erosion processes on the basis of which authorities should take the necessary steps to prevent its negative impacts on this watershed and interconnected coastal dynamics. In this regard, a number of recommended risk management strategies are being recommended for national authorities. These include the need to support effective soil conservation practices, strengthening of erosion control measures, and regular monitoring of coastal water quality for quality assurance purposes. Nowadays, novel remote sensing technologies, such as UAV-borne LiDAR and multispectral imaging of this watershed, are able to identify and quantify extremely high-resolution dynamics that can assist further risk management processes by providing a more accurate erodibility estimates and coastal water quality characterization at centimeter scale pixel resolution.

Acknowledgements

AZ acknowledges the assistance of Andrea Muscat for the provision of rain records; and that of Ines Felix for the provision of vector data.

Author details

Charles Galdies[1*], Amy Zammit[1] and Adam Gauci[2]

1 Institute of Earth Systems, University of Malta, Msida, Malta

2 Faculty of Science, University of Malta, Msida, Malta

*Address all correspondence to: charles.galdies@um.edu.mt

References

[1] Kuhwald M, Busche F, Saggau P, Duttmann R. Is soil loss due to crop harvesting the most disregarded soil erosion process? A review of harvest erosion. Soil & Tillage Research. 2002; **215**:105213

[2] Hou T, Filley TR, Tong Y, Abban B, Singh S, Papanicolaou AN, et al. Tillage-induced surface soil roughness controls the chemistry and physics of eroded particles at early Erosion stage. Soil & Tillage Research. 2021;**207**:104807

[3] Sobel R, Kiaghadi A, Rifai HS. Modeling water quality impacts from hurricanes and extreme weather events in urban coastal systems using Sentinel-2 spectral data. Environmental Monitoring and Assessment. 2020;**192**:307

[4] ERA. L-inhawi tar-Ramla. Natura 2000 Management Plan (SAC); 2014. p. 166

[5] MEPA. First Water Catchment Management Plan for the Maltese Islands; 2011. p. 147

[6] Cassar M. A project for the integrated management of protected coastal areas in Malta. Journal of Coastal Conservation. 2003;**9**:73-80

[7] ERA. [Internet]. 2017. Available from: https://era.org.mt/topic/natura-2000-datasheets-maps/ [Accessed: March 15, 2023]

[8] Scerri S. Geo-Environmental Survey of Ramla Bay – Gozo. Malta: Gaia Foundation; 2003

[9] Galdies C. Malta's Climate Anomaly Trends and Possible Related Socio-Economic Impacts. Malta: National Statistics Office; 2011. p. 45

[10] ERDF 156 Data. Developing National Environmental Monitoring Infrastructure and Capacity, Malta Environment and Planning Authority; 2013

[11] Conrad O. Tool Basic Terrain Analysis. SAGA-GIS Tool Library Documentation (v7.1.1) [Internet]. 2005. Available from: http://www.saga-gis.org/saga_tool_doc/7.7.1/ta_compound.html [Accessed: March 15, 2023]

[12] Conrad O, Olaya V. [Internet]. 2004a. Available from: https://saga-gis.sourceforge.io/saga_tool_doc/7.7.0/ta_channels.html [Accessed: March 15, 2023]

[13] Horton RE. Erosional development of streams and their drainage basins: Hydrophysical approach to quantitative morphology. The Geological Society of America. 1945;**56**(3):275-370. DOI: 10.1130/0016-7606(1945)56[275, EDOSAT]2.0.CO;2

[14] Strahler A. Quantitative geomorphology of drainage basins and channel networks. In: Chow V, editor. Handbook of Applied Hydrology. New York: McGraw Hill; 1964. pp. 439-476

[15] Riley SJ, De Gloria SD, Elliot R. A terrain ruggedness that quantifies topographic heterogeneity. Intermountain Journal of Science. 1999;**5** (1–4):23-27

[16] Galdies C, Azzopardi D, Sacco A. Estimates of soil erosion rates in a principal watershed in Gozo, Malta under current and future climatic conditions. Catena. 2022;**210**:105900. DOI: 10.1016/j.catena.2021.105900

[17] Ruddick K, Brockmann C, Doerffe R, Lee Z, Brotas V, Fomferra N, et al. The Coastcolour project regional algorithm

round robin exercise. In: Frouin RJ, Yoo HR, Won J-S, Feng A, editors. Remote Sensing of the Coastal Ocean, Land, and Atmosphere Environment. Proc. of SPIE 2010. Vol. 7858. 2010. p. 785807; CCC code: 0277-786X/10/$18. DOI: 10.1117/12.869506

[18] CoastColour. CoastColour—Products. [Internet] 2023. Available from: https://www.coastcolour.org/products.html [Accessed: March 15, 2023]

[19] Galdies C, Vella K. Future impacts on Malta's agriculture based on multi-model results from IPCC's CMIP5 climate change models. In: Castro PC, Azul AM, Leal Filho W, Azeitero UM, editors. Climate Change-Resilient Agriculture and Agroforestry – Ecosystem Services and Sustainability. Chapter 8. Cham: Springer; 2019. DOI: 10.1007/978-3-319-75004-0_8

Chapter 6

Erosion Control at Downstream of Reservoir Using In-stream Weirs

Yaoxin Zhang, Yafei Jia, Keh-Chia Yeh and Chung-Ta Liao

Abstract

As low-head hydraulic structures, instream weirs are built across rivers to control the upstream water surface elevation and the downstream flow conditions. This chapter presents a study of erosion control using instream weirs at downstream of a reservoir; JiJi Weir was built across the longest river in Taiwan, Chuoshui Creek, a mountainous river with steep slopes. Due to the easy-to-be-eroded fine lithology layers of mud, shiver, and sandstones on channel bed, the downstream of JiJi Weir had suffered from severe channel incision and head-cut development problems, which greatly threatens the integrity of the dam. To protect the JiJi Weir and its downstream channel from serious channel erosions, the Water Resources Agency (WRA) of Taiwan proposed erosion control plans that multiple instream weir structures were to be installed along the downstream channel of JiJi Weir. A three-dimensional (3D) numerical model, CCHE3D model with capabilities of simulating bedrock erosions, was used to evaluate those erosion control plans and thus explore for the optimal design.

Keywords: instream-weir, bedrock erosion, channel incision, erosion control, three-dimensional model

1. Introduction

JiJi Weir was built across Chuoshui River, the longest river (178.6 km) in Taiwan, originating from the central mountains and flowing into the East China Sea, covering a watershed about 4323 km^2. A shallow reservoir was formed due to the construction of the weir. The precipitation and geology dominate the flow and sedimentation processes in the channel and the reservoir. Particularly, during typhoon seasons, floods with high discharges will carry a large amount of sediments flowing through the river channel. Shortly after the weir construction, a pebble-boulder-sized sedimentation layer covering the channel bed was mobilized by the initial clear water releases, and the soft rocks beneath this sedimentation layer were exposed. Within the 6.5 km region downstream of the dam, the channel bed now is featured with erodible lithology layers of mud, shiver, and sandstones, resulted from severe channel incisions and a head-cut development. According to the comparisons of historical longitudinal profiles in **Figure 1**, the vertical incision of the channel thalweg from 2002 to 2014 is

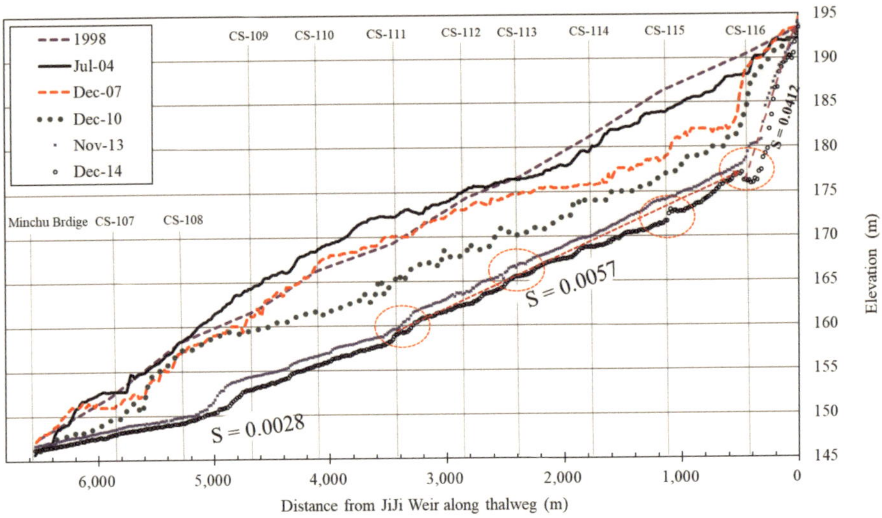

Figure 1.
Historical longitudinal profiles along thalweg (S is bed slope; CS denotes "cross section").

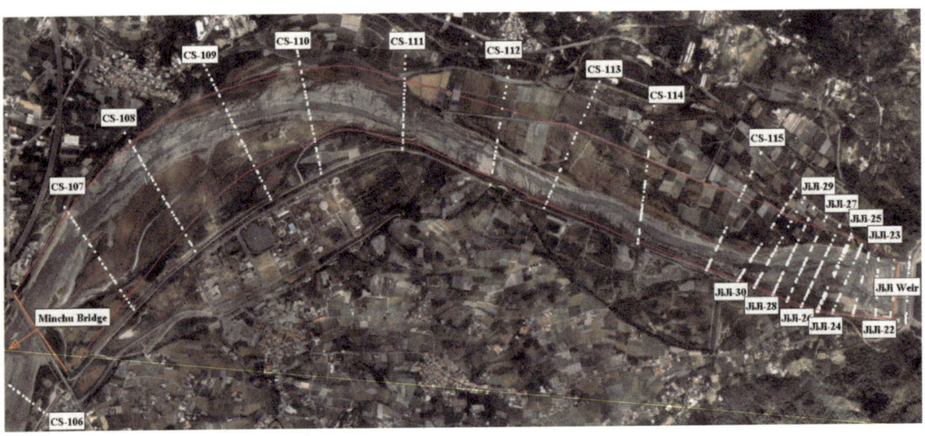

Figure 2.
Study domain with survey cross sections.

about 10 m (please refer to **Figure 2** for the locations of those measured cross sections from CS-106 to CS-116). Looking downstream, **Figure 3** shows the incised channel near the JiJi Weir. The white dash line indicates the flat channel bed before the incision. The head-cut was still actively developing and migrating, which has threatened the integrity of the dam (**Figure 4**).

To prevent this channel from further erosion and protect the JiJi Weir Dam, the Water Resources Agency (WRA) of Taiwan has proposed several erosion control plans [1, 2]. Several weir structures and channel widening were proposed to promote sediment deposition and lower the water surface level for flood protection. This study is to provide alternative plans optimal for erosional control and flood protection by using a 3D numerical model, CCHE3D [3], with capabilities in simulating bedrock erosions with lateral erosions.

Figure 3.
Bedrock channel (photo by Zhang, Y., 2015.11.2; look toward downstream).

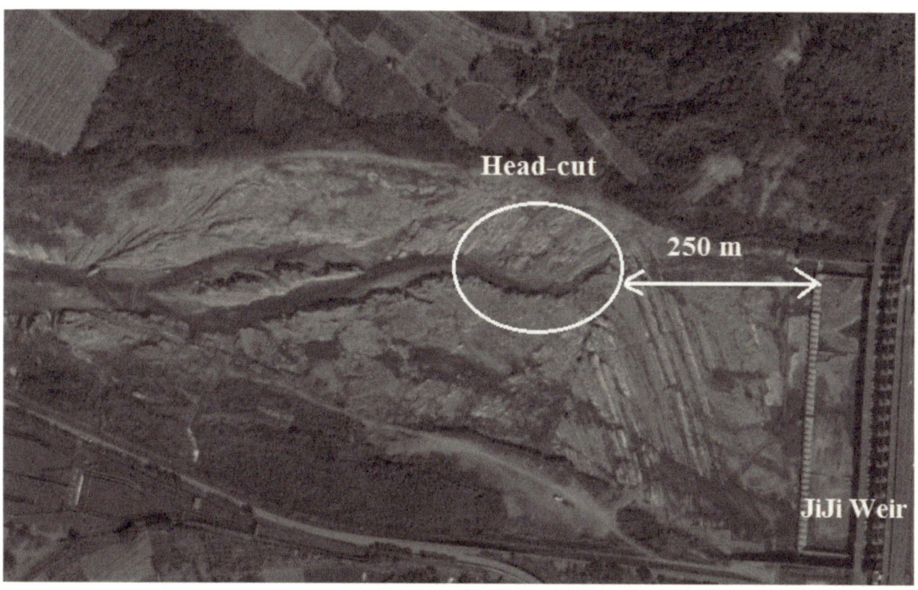

Figure 4.
Head-cut development at downstream of JiJi Weir (satellite image, 2014.11.23).

The bed rock erosion is closely related to channel lithology, tectonics, and geomorphology properties, contributed significantly by complicated weathering, plucking, abrasion (due to both bed load and suspended load), cavitation, and dissolution processes [4], which is distinguished from river sediment transport and morphologic processes on alluvial beds. Extensive research studies have been carried out for attempting to describe these complex physical mechanisms of bedrock erosions mathematically. For examples, Annandale [5] proposed a conceptual framework to correlate the flow energy to the earth mass erodibility by introducing the erodibility index, which characterizes the capability of earth materials for resisting erosions. Whipple and Tucker [6] considered the stream power to dominate the bedrock erosions. Whipple *et al.* [4] developed the conceptual models for plucking, abrasion, and cavitation in terms of the relationship between the bed shear stress and erosion rate by studying the field erosion processes. Sklar and Dietrich [7] identified the sediment supply as an important factor for bedrock erosions due to abrasion. Sklar and Dietrich [8] proposed a mechanistic bedrock erosion model by saltating bed load based on their work in 2001. Turowski *et al.* [9] extended and improved the Sklar's approach with more detailed covering effect. Stock *et al.* [10] measured rock erosion rate in several field sites including river in the United States and Taiwan. Lamp *et al.* [11] extended Sklar and Dietrich [8]'s work further by considering the impacts from not only the bed load but also the suspended load.

From the previous studies, the fluvial impacting factors for the bedrock erosions were identified as follows: bedrock erodibility (strength), stream power, shear stress, sediment supply, and grain size. Accordingly, two types of models [12], namely the hydraulic scour model (stream power-based [5, 13]) and the abrasion scour model [8] have been used in the applications with numerical simulations. Jia *et al.* [14] incorporated the abrasion-based bedrock erosion model into CCHE2D model [15] to study the channel incisions at downstream of JiJi Weir. Lai *et al.* [12] proposed a hybrid bedrock erosion model by combining the abrasion-based model and the stream-power based model. Liao *et al.* [16] implemented a stream-power-based soft bedrock erosion model into a 2D mobile-bed model EFA (Explicit Finite Analytical) to study the channel morphological process in an uplifted reach of Ta-An River by an earthquake in Taiwan. In addition to a 2D bedrock erosion model, Jia and Zhang [1, 2, 17] first developed a 3D bedrock erosion model by extending Liao *et al.* [16]'s method to CCHE3D [3] to study the channel incision problems in Taiwan.

This chapter is to present the development and applications of the CCHE3D bedrock erosion module to study the channel incisions and erosion control problems in the downstream channel of the JiJi Weir. The numerical model was calibrated and validated using field data. Based on the evaluations of the multiple alternatives, an optimal control plan, involving multiple weir constructions with the side channel excavations as well, was proposed to control the head-cut development soft rock and the channel incision.

2. Numerical model

CCHE3D model [3] is a Finite Element Method (FEM) model based on a partially staggered 3D sigma layered mesh system, which consists of multiple layered structured meshes. It was developed for 3D numerical simulation and analysis for free surface turbulent flows in rivers, lakes, reservoirs, and estuaries and associated processes, such as sediment transport [18], morphological changes [17], heat transfer [19], pollutant transport and water quality evaluation [20], etc. Since it can provide more accurate and detailed local flow fields around in-stream structures [18] than 2D depth averaged models, CCHE3D model was selected to evaluate the erosion control plans with multiple weir structures installed in the study channel.

2.1 3D RANS model

The full 3D Reynolds-averaged Navier-Stoker (RANS) equations are solved in CCHE3D model, which are simply listed in the form of index notations for Cartesian coordinates as follows:

$$\frac{\partial u_i}{\partial x_i} = 0 \tag{1}$$

$$\frac{\partial u_i}{\partial t} + u_j \frac{\partial u_i}{\partial x_j} = -\frac{1}{\rho}\frac{\partial p}{\partial x_j} + \frac{\partial}{\partial x_j}\left(\nu \frac{\partial u_i}{\partial x_j} - \overline{u_i' u_j'}\right) + f_i \tag{2}$$

where u_i = Reynolds-averaged velocities defined at x_i $(i, j = 1, 2, 3)$; t = time ; ρ is water density; p is pressure; ν is kinematic viscosity; $-\overline{u_i' u_j'}$ = Reynolds stress; and, f_i represents the body force.

For surface flows, the free surface kinematic equation is applied:

$$\frac{\partial \eta}{\partial t} + u_\eta \frac{\partial \eta}{\partial x} + v_\eta \frac{\partial \eta}{\partial y} - w_\eta = 0 \tag{3}$$

where η is water surface elevation; and (u_η, v_η, w_η) denotes velocity at water surface.

In CCHE3D model, a partially staggered stencil and a velocity correction algorithm are used for solving the momentum and continuity equation. Several turbulence closure schemes including the zero equation models (parabolic, mixing length, and wind-induced) and the $k-\varepsilon$ models are provided. More details of CCHE3D model can be found in Ref. [3].

2.2 3D sediment transport model

Sediment transport is one of the most complex and least understood phenomena in nature. In 3D, sediment particles' movements are highly affected by vertical motion of fluid flows in addition to horizontal movements. This is particularly true in the vicinity of hydraulic structures where the flow impacts on the solid walls of the structures (bridge pier and abutment, for instance). If the structures are very large (dam), the fluid flows in an open channel are forced to change their speed and direction near structures in order to pass through them, the sediment transport capacity due to this impact is adjusted significantly causing localized scouring and deposition over the sediment bed. In CCHE3D model, in addition to general sediment transport capabilities, special sediment transport features, such as local scouring around structures and channel head-cut migration have been developed as well.

In CCHE3D model, the 3D convection-diffusion equation for the suspended sediment is solved as follows:

$$\frac{\partial C}{\partial t} + u\frac{\partial C}{\partial x} + v\frac{\partial C}{\partial y} + (w - \omega_s)\frac{\partial C}{\partial z} - \frac{\partial}{\partial x}\left[\frac{\nu_t}{\sigma_s}\frac{\partial C}{\partial x}\right] - \frac{\partial}{\partial y}\left[\frac{\nu_t}{\sigma_s}\frac{\partial C}{\partial y}\right] - \frac{\partial}{\partial z}\left[\frac{\nu_t}{\sigma_s}\frac{\partial C}{\partial z}\right] = ST \tag{4}$$

where C is the suspended sediment concentration; u, v, and w are velocity components (m/s); ω^s is the sediment settling velocity; ν_t is the eddy viscosity (m^2/s); ST is the source term. and σ^s is the Schmidt number to convert the turbulence eddy viscosity to eddy diffusivity for suspended sediment.

At the free surface, the vertical sediment flux is zero, so the gravity effects $\omega_s C$balance the diffusion effects $\varepsilon_s \frac{\partial C}{\partial z}$, and the following condition is applied:

$$\omega_s C + \varepsilon_s \frac{\partial C}{\partial z} = 0 \tag{5}$$

At the bottom, the following condition is applied:

$$\omega_s C + \varepsilon_s \frac{\partial C}{\partial z} = D_b - E_b \tag{6}$$

where ω_s is settling velocity (m/s); $\varepsilon_s = \nu_t/\sigma_c$ is diffusion coefficient for sediment; D_b and E_b (kg/m^2/s) are deposition rate and erosion (re-suspension) rate at bottom, respectively.

Following the non-equilibrium transport approach ($q_{bk} \neq q_{b_*k}$) proposed by Wu [21], the bedload transport rate is governed by

$$\frac{\partial(\delta \bar{c}_{bk})}{\partial t} + \frac{\partial q_{bkx}}{\partial x} + \frac{\partial q_{bkx}}{\partial y} = -\frac{1}{L_b}\left(q_{bk} - q_{b_*k}\right) \tag{7}$$

where q_{bk} is the bedload transport rate for the k_{th} size class, q_{bkx} and q_{bky} are the component in x and y directions, δ is the bedload layer thickness, $\bar{c}_{bk}$ is the bedload concentration and L_b is the bedload adaptation length; q_{b_*k} is the bedload sediment transport capacity for equilibrium transport conditions, which can be estimated using empirical transport formulas.

2.3 Bedrock erosion model

According to previous studies, the bedrock erodibility (strength), stream power, shear stress, sediment supply, and grain size were identified as important impacting factors on bedrock erosion rate, which lead to two popular bedrock erosion mechanisms, plucking and abrasion, widely used in the numerical models [1, 12, 14, 16]. The first one is corresponding to the so-called stream power-based method that the bedrock erosion rate is considered as a function of the stream power, and the varying shear stress causes the hydraulic scouring on soft bedrock [4–6, 13], and the other one is the abrasion-based method, which emphasizes on the important role of the sediment supply (both bedload and suspended load) by considering the eroding and shielding effects of sediment on bedrock [7, 8, 11].

Whipple *et al.* [4] observed in the field that for well-joined rocks with fractures and bedding planes, the plucking is the dominant erosion process, while the abrasion process dominates for rocks with smooth and polished surfaces but with ripples, flutes, and potholes prominently developed. All the aforementioned bedrock erosion models oversimplified and conceptualized the complicated bedrock erosion processes in nature. For any particular mountainous river, these natural processes cannot be separated and modeled accurately using one method. Practically, however, a certain dominant process has to be selected to represent all erosion mechanisms for the modeling purpose. For the downstream channel of JiJi Weir, since the measured bedrock erodibility is available, it is assumed that the plucking is dominating in this reach, and the stream power method [16] is selected for current study.

In the stream power method, the bedrock erosion rate E is only related to the rock erodibility index [5] and the flow stream power, which is proportional to bed shear stress, as described in Eqs. (8, 9):

$$E = K_s U \left(\frac{P}{P_{cm}} - 1 \right)^c = K_s U \left(\frac{\tau U}{P_{cm}} - 1 \right)^c \tag{8}$$

$$P_{cm} = a K_h^b \tag{9}$$

where K_s is non-dimensional coefficient; U is depth-averaged velocity of flow (m/s); P is stream power of flow (kW/m^2), $P = \tau U$, τ is shear stress (N/m^2); P_{cm} is critical stream power (kW/m^2); K_h is the bedrock erodibility index defined as the capability of earth materials for resisting erosion, which is correlated empirically to the stream power and obtained based on field and laboratory studies; and, a, b, and c are site-specific calibrated parameters.

The above stream power method is further improved to take into account lateral erosion by considering the local lateral bed slope. Thus, a factor S_b representing high slope zones is introduced as follows:

$$S_b = \max \left(\frac{S_l}{k \cdot S_R}, 1.0 \right)^r \tag{10}$$

where S_l is the local lateral bed slope computed in an element, S_R is a reference slope of the simulation area; it is currently represented by the average slope of all wet elements in a domain. The power r is empirical and needs to be calibrated. In the tests, it is found that r = 1.5~5.0. It can be seen that this factor is effective only when the local lateral bed slope is larger than the reference slope. k = 1~6 is the coefficient to adjust the reference slope, which filters out small slope area from erosion. With this factor, Eq. (8) is modified to:

$$E_{bb} = K_s U \left(\frac{P}{P_{cm}} - 1 \right)^c S_b \tag{11}$$

where E_{bb} is the erosion rate applicable to both softrock bed and bank erosion. With r = 0, Eq. (11) will convert back to Eq. (8).

2.4 Coupling of bedrock erosion model and sediment transport model

When the sediment transport is simulated, the boundary condition between the moving sediment particles and the bedrock has to be treated. Sediment particles can deposit over the rock surface and form a deposition layer. A concept of a sediment mixing layer over the softrock surface is adopted. If the thickness of the sediment layer is large, no rock erosion is calculated. If no sediment deposition exists on the bed, the proposed stream power method is used for the rock bed erosion. If the thickness of the sediment layer is within a criterion (mixing layer thickness), the stream-power-induced erosion would be applied at a reduced rate, proportional to the thickness of the mixing layer. The net change rate of the mixing layer is the combined rates of rock erosion and sediment deposition.

There is insufficient knowledge on the interactive and coupling mechanism between the bedrock erosion and the sediment transport. This simple coupling basically considers the shielding effects of the sediments on bedrock, similar to that of the abrasion model, which is concetually reasonable and has been proved and validated in previous studies [7, 8, 11]. However, the actual interactions between the bedrock erosion process and the sediment transport process are much more complicated in nature.

3. Application to downstream of JiJi Weir

In this study, CCHE3D bedrock erosion model was used to simulate the aforementioned channel incision process. The validated model was then used to evaluate erosion control plans. The study domain is a 6.5 km reach from the JiJi Weir to the downstream Minchu Bridge, where there are measured cross sections available: CS-106 to CS-116, and JiJi-30 to JiJi-22 (**Figure 2**).

3.1 Model setup

Based on the DEM data of 2013 and 2014, several structured meshes covering the study domain from JiJi Weir to Minchu Bridge were generated, which embed the combinations of the planned weir structures and lateral channel excavations.

As shown in **Figure 5**, zones representing erodible bedrocks, alluvium, structures (non-erodible), have been identified and embedded in simulation domain based on the satellite image of 2014. The erodible bedrock zone resulted from the channel incisions and head-cut development, where both the bedrock erosion and sediment transport simulations are to be conducted. The alluvial zone is covered with pebbles

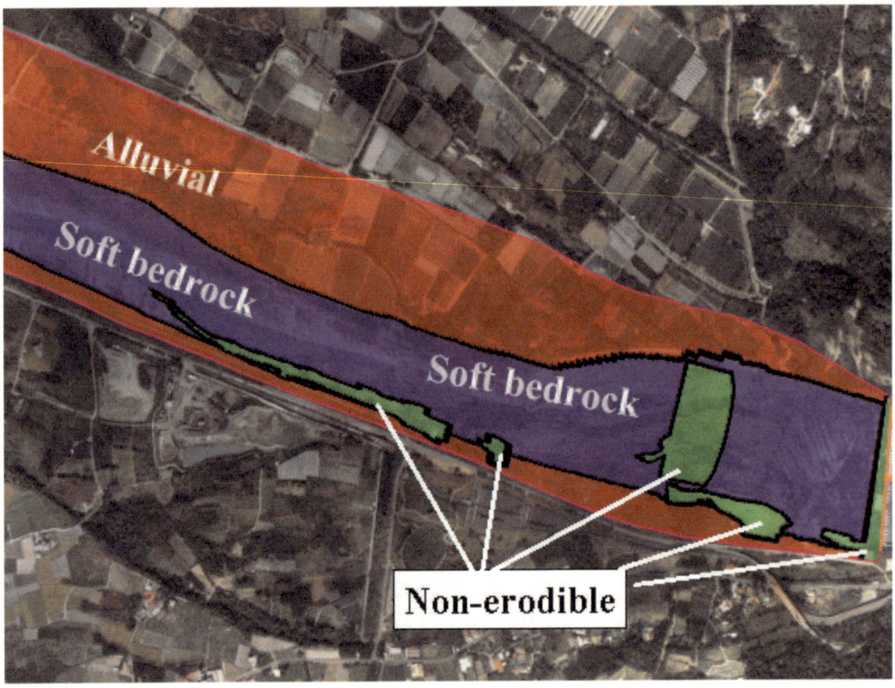

Figure 5.
Erodible bedrock zone.

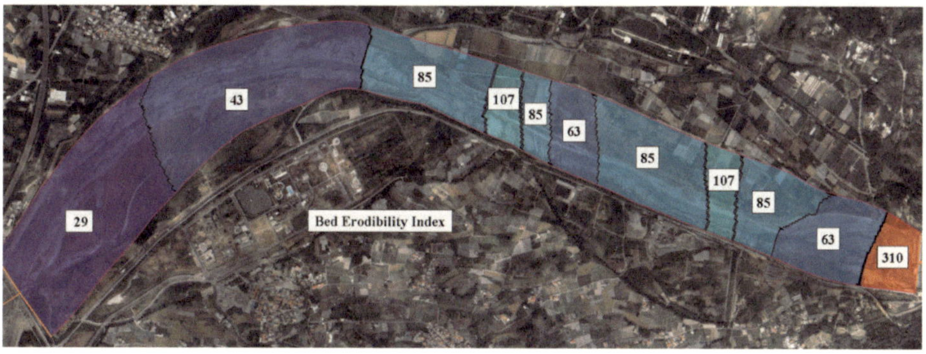

Figure 6.
Bedrock erodibility index.

and gravels, and no bedrock erosion simulation will be applied. The non-erodible zone represents the weir structures and other bank/bed protection measures (i.e., concrete blocks) and is assumed non-erodible during the simulations. Therefore, the distribution of non-erodible zones changes with the simulated erosion control plans.

Figure 6 shows the measured bedrock erodibility index distribution in 2014. The whole domain was divided into 11 zones with different erodibility index, k_h, varying from 310 (high rock strength) to 29 (low rock strength) [2]. The aforementioned channel bed properties (**Figures 5** and **6**) are applied to all cases. Adjustments are made accordingly for the installations of weir structures in different erosion control plans.

3.2 Model calibration and validation

The site-specific parameters (a, b, and c) in Equation (8) and the lateral slope parameters r and k in Equation (11) need to be calibrated before applications. In this

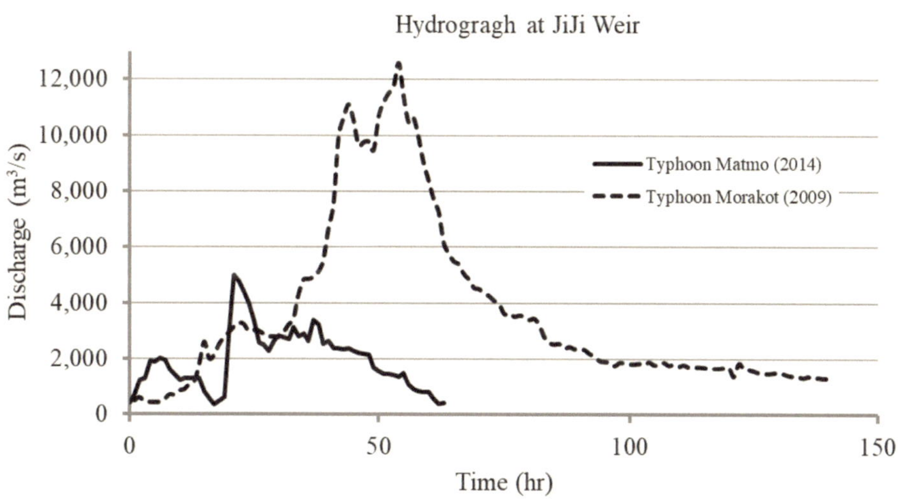

Figure 7.
Flood hydrograph of Typhoons Morakot (8/2009) and Matmo (7/2014).

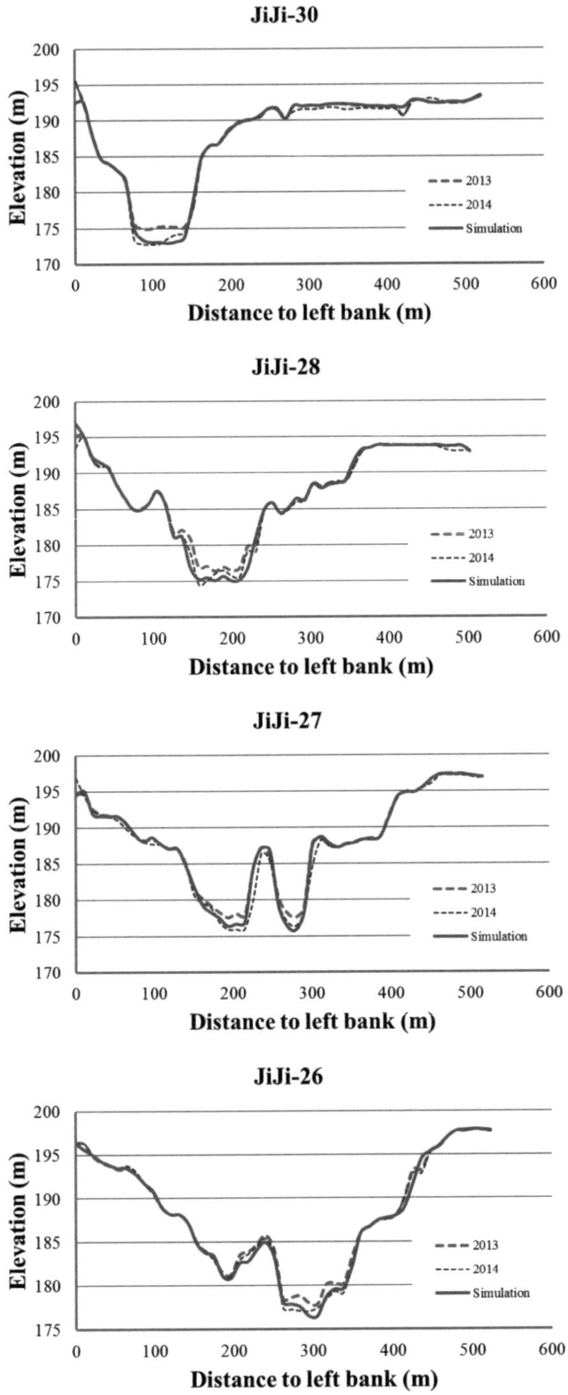

Figure 8.
Comparisons of cross-sectional profiles for P0-1 with Typhoon Matmo.

study, two cases without any control structures, P0-1 with Typhoon Marmo and P0-2 with Typhoon Morakot, are used to calibrate these parameters and validate the bedrock erosion model, respectively.

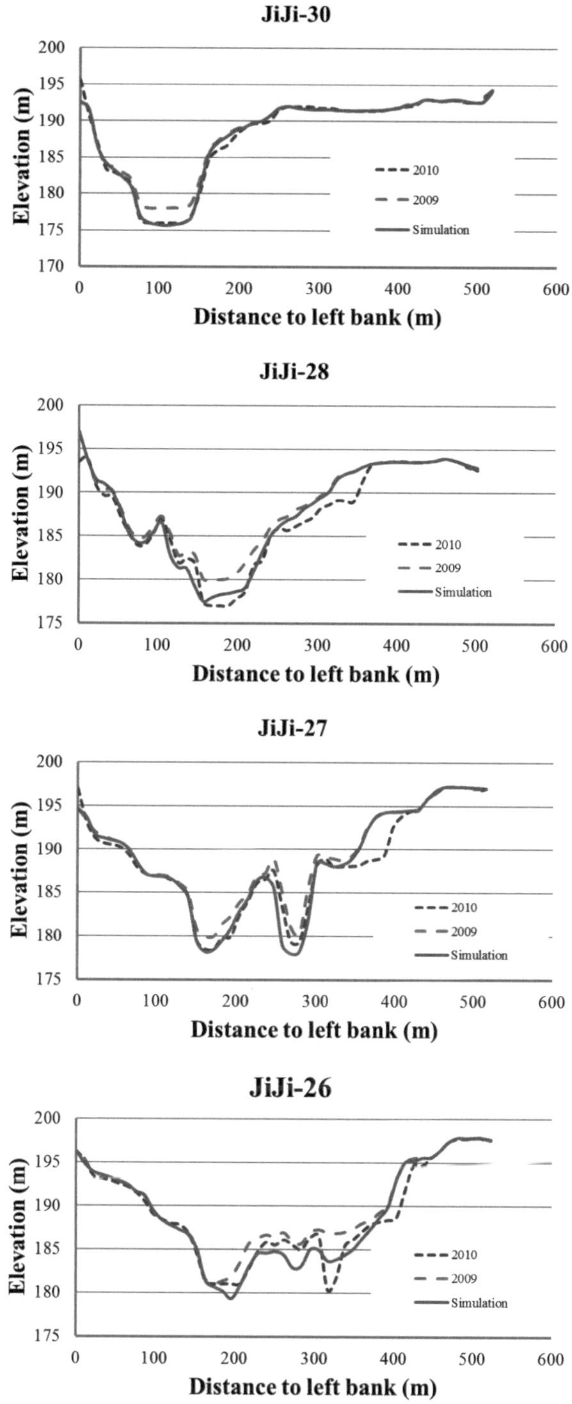

Figure 9.
Comparisons of cross-sectional profiles for Po-2 with Typhoon Morakot.

In the calibration step, the parameters are adjusted to obtain the best fit of the simulations to the data; while in the validation step, no parameters are changed. **Figure 7** shows the hydrographs of these two typhoons, which are considered as the major hydrological events causing significant bed morphological changes in this channel. Due to its high peak discharge (Q = 12,600 m^3/s), Typhoon Morakot was selected for all other erosion control plans.

Through the numerical calibration tests, the parameter set for the stream power method was found to be: a = 0.005, b = 0.75, and c = 0.2. For the lateral erosion effects, the slope exponential parameter r = 1 and the slope scale factor k = 6 were used. **Figures 8** and **9** compare the profiles at the selected cross sections for P0-1 with Typhoon Matmo and P0-2 with Typhoon Morakot, respectively. In general, good agreements between the measurements and the simulations were observed for both cases at those cross sections, especially for channel incisions. As for the lateral erosion observed at sections JiJi-28, 27, and 26, although more discrepancies exist, the improved stream power method with considering the slope effects (Equation 7) has proved its capability of capturing these lateral erosion phenomena in the complicated bedrock erosion processes.

3.3 Erosion control plans

Since 2007, WRA has proposed a few erosion control plans attempting to stop the downstream channel of JiJi Weir from incisions and head-cut development. These erosion control plans included multiple in-stream weir structures, lateral channel excavations, and other engineering measures, such as concrete blocks for bank protections, gabions, etc. With the calibrated and validated parameters, CCHE3D bedrock erosion model was used to evaluate all the erosion control plans with different combinations of control structures [2]. According to the numerical simulations, one plan with three weirs at JiJi-22, 25, and 26 and side excavations from JiJi-27 to CS-112 was confirmed as the most effective in the erosion reduction among all proposed erosion control plans. Based on the numerical simulations of these erosion control plans, optimal design was explored.

In the historical channel profiles along thalweg of the downstream channel from 1998 to 2014 as shown in **Figure 1**, the bed slope is 0.0412 near the head-cut reach, 0.0057 in the incision reach, and 0.0028 in the transition reach. The head-cut reach and the incision reach are of bare rock channel, actively eroded; the transition reach is sometimes partially covered with sediments, showing alluvial river morphologic and sediment transport features, and thus considered to be more stable. In this study, it is assumed that the channel will be stabilized if the bed slope can be reduced to 0.0028 approximately by installing erosion control weirs in the channel.

Since the 10 m deep head-cut has reached closer to the JiJi Dam, seriously threatening the safety of the dam structure (**Figure 3**), the protection of the upstream head-cut zone (from JiJi-30 to JiJi-22) is considered as the first priority. The high weir structure at JiJi-25 is capable of significantly reducing the bed rock erosion in the reach from JiJi-26 to JiJi-22. As for the reach from JiJi-30 to JiJi-26, a weir structure is planned at CS-115, and the small reservoir behind this weir structure is designed to slow down the flow and thus reduce the erosion.

For downstream of the head-cut zone, the deep incised main channel (from CS-109 to JiJi-26) is relatively narrow, the water surface elevation in this reach will increase during floods to endanger the embankment of the left bank. Channel lateral excavations are proposed in such a way that the thalweg is kept and the widening lateral excavation (150 m) is on the right floodplain of the channel from JiJi-27 to CS-108.5. Hydrologic data indicates that the flow discharge in the channel is less than Q = 1000 m^3/s for most time of the year, and the flows in this range are confined in the main channel. The water depth for this discharge is about 4 m. If no in-stream weir structures are built, the proposed lateral excavations will still allow the water to flow in the deep channel for the most of time of a year, but the water will divert to the widened areas in the flood seasons. The widened channel will reduce the main flow velocity and water surface elevation, which is beneficial to bank protection during floods in addition to promote sediment deposition.

To enhance the erosion reduction effects, two additional low-headed weirs are suggested to be installed at CS-111 and CS-113 in this reach to further control the flow and erosion. To prevent the development of a second head-cut between CS-108 and CS-109 (see **Figure 1**), another low-headed weir structure is proposed as well to be installed at CS-108.5.

The heights of the four weirs at CS-108.5, CS-111, CS-113, and CS-115 are determined in such a way that the small pools formed behind the weirs can approximately protect half of the reach between weirs (slope equal to zero). As illustrated in **Figure 10**, the bed slope of the reach between Weir-A and B is S_0, the pool behind Weir-A can affect an area about half of the reach ($L/2$). Sediments would fill up the pool behind Weir-A because the surface slope is significantly reduced. Before the deposition filled the channel segment, the bed slope behind Weir-B is larger than S_d ($= S_0/2$), and it reaches to S_d after the segment is filled up, which is the highest slope possible in the design channel. The new established bed slope would be about half of the initial slope S_0. The protected channel would be stable and filled with sediment. According to this idea, the top elevations of the five weirs are determined as indicated in **Table 1**.

Figure 11 shows the initial setup for the optimal design. The excavation is denoted in the white polygon area, while six weir strucrures installed at JiJi-25 (CS-116), JiJi-26, CS-115, CS-113, CS-111, and CS-108.5) as non-erodible zones. For the optimal design case simulations, the calibrated parameter sets and the bedrock erodibity index

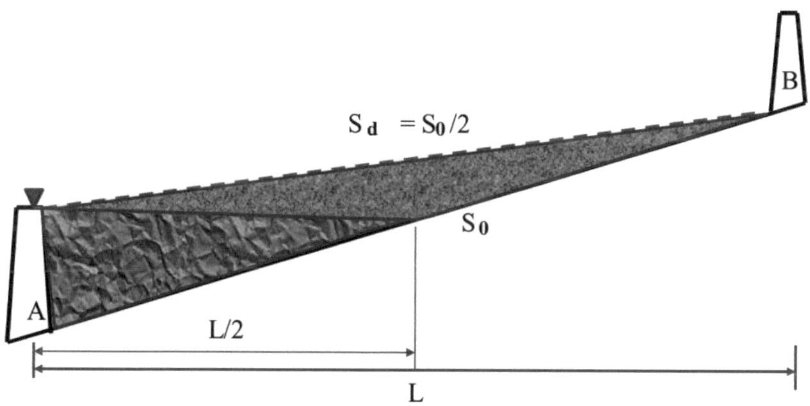

Figure 10.
Sketch of determination of weir height.

Weirs	CS-108.5	CS-111	CS-113	CS-115	JiJi-26	JiJi-25
Elevation (m)	155.5	163	170	177	181	188

Table 1.
Top elevations of weir structures.

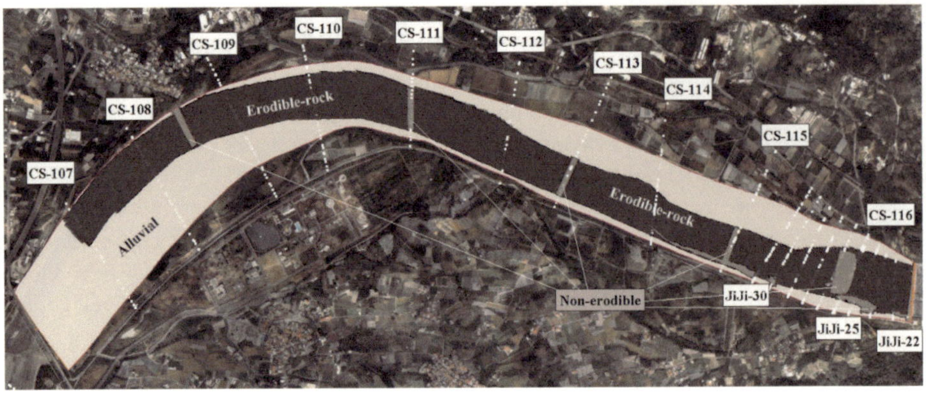

Figure 11.
Model setup for optimal design.

remain the same. The bed property zone is adjusted if the excavation cuts into the alluvium and exposes the soft rock underneath.

In current sediment transport simulations, both the suspended load and the bedload transport are considered with three representative size diameters: 0.0081, 0.03086, and 0.3006 m. Sediment boundary condition is obtained from the measured concentration in a detention pool nearby JiJi Weir during typhoon season, initial conditions and other boundary conditions are the same as the cases of stream power erosion simulations. This simply coupled sediment transport and stream-power-based erosion model was calibrated using the P0-1 conditions and then applied to sedimentation simulations in the optimal design study. The calibrated model could capture the main trend of the channel incision particularly in the reach from the JiJi Dam to CS111 under the P0-1 condition.

The simulation results of the optimal design were compared with those of the cases without any structures. **Figures 12** and **13** show the longitudinal profile of bed changes. In the first 1.2 km reach, where two high weirs at JiJi-25 and JiJi-26 are

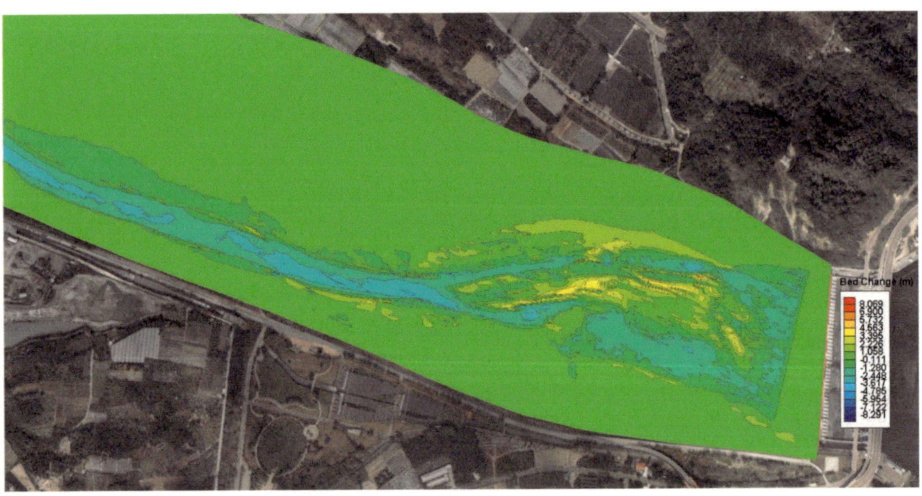

Figure 12.
Simulated erosion patterns for the case without structures.

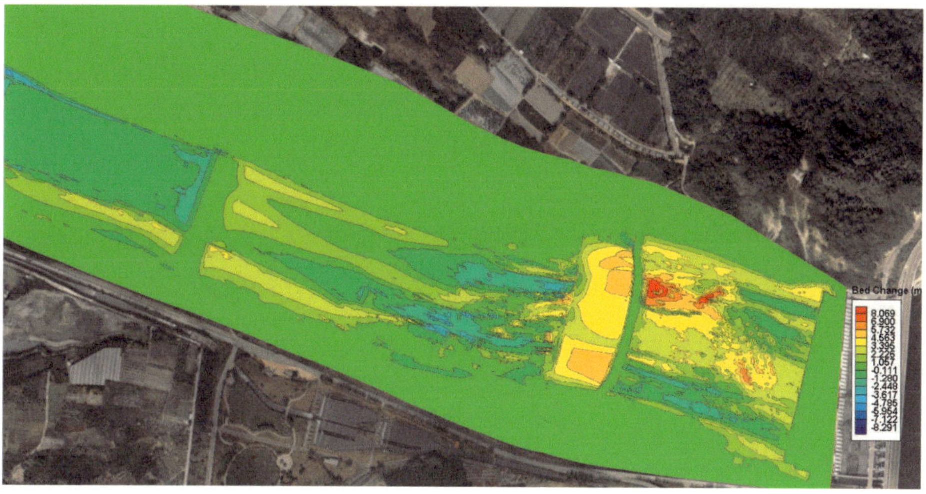

Figure 13
Simulated erosion patterns for the case with the optimal design.

located, much more depositions were observed in the optimal plan, while the erosions are dominant for the case without structures. Further downstream, in general, the erosions are dominant for both cases, except for small regions upstream of the weir structures, where the water surface is increased, and the velocity is slower . The bed change pattern around all the weir structures is similar: there are depositions at upstream but erosions at downstream, which is expected. **Figure 14** shows the bed profile along the thalweg. As can be seen, in the optimal plan, the erosion along the thalweg has been reduced significantly, and the upstream reach with the first two weirs demonstrated obvious deposistion pattern. The two weir structures at JiJi-25 and JiJi-26 do serve the purpose of reducing channel incision and stopping the head-cut development effectively (**Figure 15**).

Despite the simple coupling method, the coupled bedrock erosion and sediment transport model demonstrated the promoting effects of the proposed optimal design on sediment depositions to protect the channel bed from further eroding.

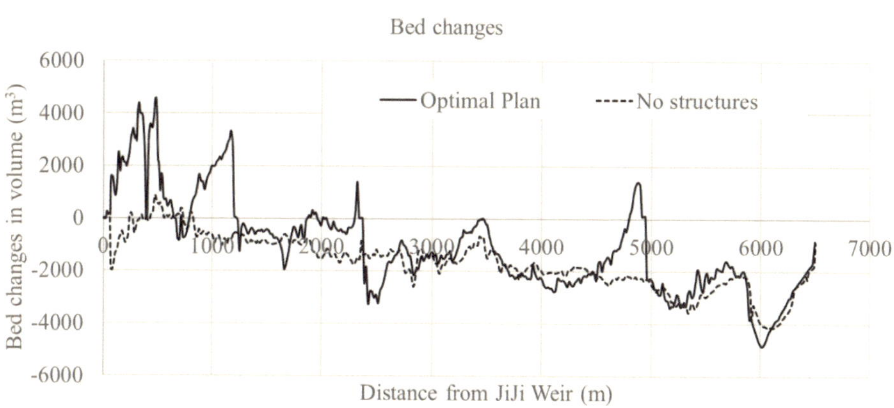

Figure 14
Profile of bed changes.

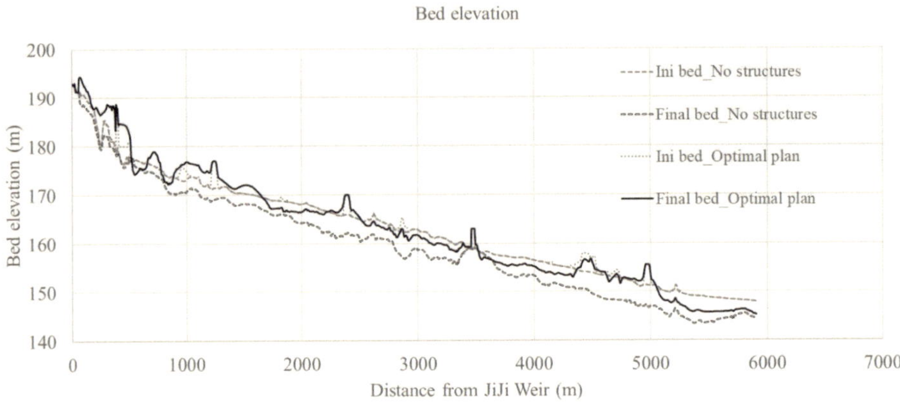

Figure 15
Profile of bed elevation along the thalweg.

3.4 Implementation of erosion control plan

The erosion control plan combined with other engineering measures (concrete blocks, gabions, and filling-ups) has been partially or completely implemented. **Figure 16** compares the aerial images of JiJi Weir in 2014 and 2021. In this 1.2 km

Figure 16
Aerial images of JiJi Weir.

reach of JiJi Weir downstream, a comb weir structure, two drop structures, and some concrete blocks were installed. The head-cut development from 2014 to 2021 seems not very significant, which demonstrated the effectiveness of the erosion control plan.

4. Conclusions

In this study, a simply coupled 3D bedrock erosion and sediment transport model, CCHE3D bedrock model, was developed and applied to the downstream channel of the JiJi Weir, Chushui River, where the serious channel incision and head-cut migration endanger the JiJi Dam. In this simple coupling model, a concept of a sediment mixing layer over the bedrock surface is adopted. The thickness of the sediment layer determines the bed change mechanism as follows: thick sediment layer leads to non-rock erosion calculation, while non-sediment layer indicates the dominant role of the bedrock erosion. Within the mixing layer thickness, the stream-power-induced erosion would be applied at a reduced rate, proportional to the thickness of the mixing layer. The net change rate of the mixing layer is the combined rates of rock erosion and sediment deposition.

The hydrologic data of Typhoon Matmo in 2014 and Typhoon Morakot in 2009 were used to calibrate and validate the site-specific parameters for the stream power method–based bedrock erosion model. This model was also applied to simulate the bed changes with designed channel erosion control structures installed. The objective of the study is to identify an optimal design to alleviate channel incision and stop the head-cut development.

Based on the evaluations of multiple erosion control plans proposed by WRA and the channel evolution analysis, an optimal erosion control plan was identified. Six weir structures at section JiJi-25, JiJi-26, CS-115, CS-113, CS-111, and CS-108.5 and the 150 m-wide lateral excavations from JiJi-27 to CS-108.5 along the right floodplain to widen the channel were proposed. The lateral excavation would keep flows in the thalweg, but divert the water onto the excavated area during floods. According to the longitudinal profiles along the thalweg (**Figure 1**), the CS-115, CS-113, CS-111, and CS-108.5 were selected as the locations for installing weir structures to control the channel bed slopes. The elevations of those weir structures were determined in such a way that the target weir protects a half of a segment between two neighboring weirs so that the bed slope of the pool between weirs could be reduced by half (**Figure 10**).

Using the proposed coupled bedrock erosion and sediment transport model, the optimal design and the case without any structures were evaluated and compared. According to the simulated results, the optimal design reduced the channel incision significantly, and the head-cut development was stopped by the two weir structures installed at JiJi-25 and JiJi-26. Sediment depositions were observed not only in the deep channel but also the excavated area. In general, the whole study reach from JiJi Weir to Minchu Bridge still demonstrated erosion pattern except for the first 1.2 km reach at upstream. Without any control structures, the channel would be further deepened, and the development of the head-cut would continue.

Acknowledgements

This work is a result of research sponsored by National Chiao Tung University of Taiwan under Research Agreement No. 300212267A and The University of Mississippi.

It is also a part of research sponsored by the USDA Agriculture Research Service under Specific Research Agreement No. 6060-13000-025-00D (monitored by the USDA-ARS National Sedimentation Laboratory) and The University of Mississippi.

Conflict of interest

The authors declare no conflict of interest.

Appendices and nomenclature

C	suspended sediment concentration
D_b	deposition rate (m^2s^{-1})
E_b	erosion rate (m^2s^{-1})
E_{bb}	erosion rate for both bedrock and bank erosion (m^2s^{-1})
K_h	bedrock erodibility index
L_b	bedload adaptation length (m)
q_b	bedload transport rate (m^2s^{-1})
S_b	bed slope factor
u	x velocity (ms^{-1})
v	y velocity (ms^{-1})
w	z velocity (ms^{-1})
U	depth-averaged velocity of flow (ms^{-1})
P	stream power of flow (kW m^{-2})
η	water surface (m)
ν	kinematic viscosity (m^2s^{-1})
δ	bedload layer thickness (m)
τ	shear stress (Pa)
ω^s	sediment settling velocity (ms^{-1})
σ^s	Schmidt number

Author details

Yaoxin Zhang[1*], Yafei Jia[1], Keh-Chia Yeh[2] and Chung-Ta Liao[2]

1 National Center for Computational Hydroscience and Engineering (NCCHE), University of Mississippi (UM), Oxford, USA

2 Natural Hazard Mitigation Research Center (NHMRC), National Chiao Tung University (NCTU), Hsin Chu, Taiwan

*Address all correspondence to: yzhang@ncche.olemiss.edu

References

[1] Jia Y, Zhang Y. Preliminary Study of Local Scouring Using CCHE3D and Its Application Potential to Softrock Erosion in Choshuixi, Taiwan. National Center for Computational Hydroscience and Engineering. USA: University of Mississippi; 2013. p. 38677

[2] Jia Y, Zhang Y. Enhancement and Application of CCHE2D/3D Soft-rock Models to Chuoshui Creek of Taiwan. National Center for Computational Hydroscience and Engineering. USA: University of Mississippi; 2015

[3] Jia Y. CCHE3D Technical Manual, National Center for Computational Hydroscience and Engineering. USA: The University of Mississippi; 2013

[4] Whipple KX, Hancock GS, Anderson RS. Riverincision into bedrock: Mechanics and relative efficacy of plucking, abrasion, and cavitation. Geological Society of American Bulletin. 2005;**112**(3):490-503

[5] Annandale GW. Erodibility. Journal of Hydraulic Research. 1995;**33**(4): 471-494

[6] Whipple KX, Tucker GE. Dynamics of the stream power river incision model: implications for height limits of mountain ranges, landscape response time scales, and research needs. Journal of Geophysical Research. 2005;**104**(B8): 17661-17674

[7] Sklar LS, Dietrich WE. Sediment and rock strength controls on river incision into bed rock, Geological Society of America. Geology. 2009;**29**(12): 1087-1090

[8] Sklar LS, Dietrich WE. A mechanistic model for river incision into bedrock by saltating bed load. Water Resources Research. 2004;**W06301**(40):1-21

[9] Turowski JM, Hovius N, Meng-Long H, Lague D, Men-Chiang C. Distribution of erosion across bedrock channels. Earth Surface Processes Landforms. 2005;**33**(3):353-363

[10] Stock JD, Montgomery DR, Dietrich WE, Sklar L. Field measurements of incision rates following bedrock exposure: Implications for process controls on the long profiles of valleys cut by rivers and debris flows. GSA Bulletin. 2005;**117**(11/12):174-194

[11] Lamb MP, Dietrich WE, Sklar LS. A model for fluvial bedrock incision by impacting suspended and bed load sediment. Journal of Geophysics Research. 2008;**113**:F03025

[12] Lai Y, Greimann B, Wu K. Soft Bedrock erosion modeling with a two-dimensional depth-averaged model. Journal of Hydraulic Engineering. 2005; **137**(8):804-814

[13] Annandale GW. Scour Technology, Mechanics, and Engineering Practice. New York: McGraw Hill; 2006

[14] Jia Y, Zhang Y, Wang S-SY. Computational Study of Softrock Erosion in a Mountain River. In: 13th Cross-Straits Symposium of Hydraulic and Water Resources. Taichung, Taiwan; 2009

[15] Jia Y, Wang SSY. Numerical model for channel flow and morphological change studies. Journal of Hydraulic Engineering. 1999;**125**(9):924-933

[16] Liao C-T, Keh-Chia K, Huang M-W. Development and application of 2-D

mobile-bed model with bedrock river evolution mechanism. Journal of Hydro-environment Research. 2013;**8**(3): 210-222

[17] Jia Y, Zhang Y. Enhancement of CCHE2D/3D Models on Softrock Erosion, Local Mesh Refinement and Their Applications to Taiwan's Rivers. National Center for Computational Hydroscience and Engineering. USA: University of Mississippi; 2014

[18] Jia Y, Scott S, Xu Y, Huang S, Wang SSY. Three-Dimensional numerical simulation and analysis of flows around a submerged weir in a channel bendway. Journal of Hydraulic Engineering. 2005;**131**(8):682-693

[19] Chao X, Jia Y, Wang SSY. 3D numerical simulation of turbulent buoyant flow and heat transport in a curved open channel. Journal of Hydraulic Engineering. 2009;**135**(7): 554-563

[20] Chao X, Jia Y, Shields FD Jr, Wang SSY, Charles M. Three-dimensional numerical modeling of water quality and sediment-associated processes with application to a Mississippi Delta Lake. Journal of Environmental Management. 2009; **91**(7):1456-1466

[21] Wu W. CCHE2D Sediment Transport Model, Technical Report No. NCCHE-TR-2001-3, NCCHE, The University of Mississippi

Chapter 7

A Study of Morphological Changes in the Coastal Areas and Offshore Islands of Sudarban Coastline Using Remote Sensing

Partha Sarathi Mahato

Abstract

The Sundarbans, located along the coastal areas of India and Bangladesh, is the largest remaining single block of mangrove forest in the world, covering approximately 1 million hectares (10,000 km^2) of the Ganges-Brahmaputra delta. This unique ecosystem is under threat from major disturbances such as sea level rise and alterations in water flows from the Himalayan headwaters. There have been very few studies on the current status and dynamics of the Sundarban's coastline. To address this knowledge gap, we conducted a study utilizing Landsat images spanning from 1975 to 2022. Our findings reveal that the rates and directions of erosion and accretion varied across the different periods. During the 1994–2005 interval, erosion reached its peak with a land loss rate of ~17 km^2 per year. However, this rate substantially declined in subsequent periods to ~8 km^2. Accretion, on the other hand, showed a rate of ~7 km^2 per year between 1975 and 1988 but declined to approximately ~6 km^2 per year between 1988 and 1994. While the accretion rate has declined in recent years, the erosion rate has remained relatively high.

Keywords: Sundarbans, Bengal delta, coastline, dynamics, satellite imagery, erosion, accretion

1. Introduction

The Sundarban area of the South Asian Region is located between 21° N to 26°30'N and 88 ° E to 92° 30'E in the Bay of Bengal. It consists of a cluster of low-lying islands covering an area of ~10,000 km^2 that were formed during the last 11,000 years. It is the second-largest river delta built by the Ganga-Bramhaputra river system driven by southwest monsoon rains. The rivers carry large amounts of sediment loads, ~10^9 t/year, from the Himalayas and upper parts of the Bengal delta. The mangrove forest of the Sundarbans is shared between India (38%) and Bangladesh (62%). The Royal Bengal Tiger, Ganga River Dolphin, and certain endangered species, such the River Terrapin, may all be found in the Subarnarekha Mangrove woods, which are rich in biodiversity. In both nations, the vital ecosystem is protected and listed as a UNESCO World Heritage Site.

The mangrove ecosystem holds immense ecological and economic importance. It is home to numerous organisms that possess substantial ecological and economic values. These ecosystems play a crucial role in supporting both terrestrial and aquatic food chains, thereby sustaining a diverse range of plant and animal species. One of the notable contributions of mangrove ecosystems is their ability to serve as natural barriers, protecting shoreline and island areas from various natural hazards such as cyclones, hurricanes, and tsunamis. By absorbing and dissipating the force of waves, they effectively mitigate coastal erosion. Moreover, mangroves act as biological filters, helping to maintain water quality by trapping sediment and nutrients, thus purifying polluted coastal waters.

Furthermore, mangroves play a vital role in maintaining the carbon balance in coastal areas. They sequester and store large amounts of carbon dioxide, contributing to climate change mitigation efforts. Additionally, these ecosystems hold significance for tourism and recreation purposes, attracting visitors who appreciate their unique beauty and ecological value [1]. Overall, the preservation and conservation of mangrove ecosystems have far-reaching ecological, economic, and societal benefits.

Mangroves demonstrate remarkable ecological stability in terms of persistence and resilience. However, they are highly sensitive to changes in hydrology. Therefore, it is essential to prioritize the protection and restoration of mangrove ecosystems. Since gaining independence, a significant number of homeless individuals have migrated and settled in the reclaimed Indian Sundarban region. Despite various efforts to safeguard mangrove resources, they face substantial anthropogenic pressure resulting from unsustainable exploitation for multiple purposes, such as wood harvesting, fodder collection, fuel extraction, and charcoal production [2]. Additionally, the conversion of forested areas into aquaculture and agricultural lands, as well as the construction of jetties and harbors to meet the demands of the growing population, further exacerbate the challenges faced by mangroves [2–4].

On an average, 5–6 cyclones annually hit the Sundarban area. Of these cyclones, two are of severe category. Among these, Amphan in 2020 had the highest impact with an estimated loss of 128 lives and > USD 13 million in damages. The diversity and extent of the Sudarban are constantly declining due to anthropogenic and natural causes. Regular monitoring of extent and quality is necessary for this area to control or even regain the loss if given constant effort for a longer period of time. Land dynamics of delta coastline are controlled by three major factors.

1. Tectonic subsidence and compaction

2. Relative sea level change and wave action

3. River sediment supply

Studies suggest the subsidence rate of the Bengal delta area is in the range of 15–50 mm annually. The exploration of oil and gas from delta, trapping of sediments due construction of reservoirs upstream, and other anthropogenic activities are considered the main cause of the subsidence of Bengal Delta. Also, the estimated rise of sea level of Bay of Bengal is at >10 mm/year which is among the world's highest (Based on global sea level data and modeling, Eriscson et al. [5] are noticeable this coastline.

The objective of this study is to observe the change of shoreline in Bay of Bengal through time and its impact on the extent of mangrove forest cover area. The rate of shoreline erosion or accretion and movement of coastline are calculated using the series of Landsat images available from 1975 to 2022. Conducting field surveys in the swampy mangrove forest can be extremely challenging due to their inaccessibility (Nandy et al. [6]). In such circumstances, remote sensing techniques have emerged as increasingly valuable tools for mapping and monitoring mangroves in a timely manner [7]. While remote sensing data cannot replace field surveys, it offers several advantages. These include synoptic coverage, the availability of free or low-cost satellite data, and repetitive coverage [8, 9].

Remote sensing allows for the collection of data over large areas, providing a broad overview of mangrove ecosystems that would be difficult to achieve solely through field surveys. Additionally, satellite data can be obtained at regular intervals, enabling the monitoring of changes and trends in mangrove extent and condition over time. The availability of free or affordable satellite data further enhances the accessibility and affordability of remote sensing technology for mangrove monitoring purposes. While remote sensing data is a valuable tool, it is important to note that it should be complemented with field surveys to validate and ground truth the remote sensing results. By integrating remote sensing with field-based data collection, a comprehensive and accurate understanding of mangrove ecosystems can be achieved.

Allison [10] mentioned the Bengal delta is in a net erosional state at a rate of ~1.9 km^2 per year and the coastline retreat of ~3–4 km in some areas of the western edge since 1792 (~21 m $year^{-1}$). Some studies mention the accretion rate as ~7 km^2 $year^{-1}$ along the river mouth regions [11].

2. Study area

The coastline along Sundarban area, unlike adjacent areas, is not restricted by embankments, making it a preferred study area to observe the impact of sea level rise and coastal erosion on the shoreline. Anthropogenic activities, such as fishing, hunting, and resource harvesting, occurs inside the inland river channel and their subsidiaries and inland forest area. Excessive humidity prevails throughout the year. The monsoon occurs in the months of June to September with 2500 to 3000 mm of rain fall annually. In the summer temperature ranges between 25 to 35°C, while winter has a temperature range of 12 to 24°C.

Tidal levels in the area exhibit seasonal variations, ranging from 4 to 6.5 meters. Additionally, the pH of the water fluctuates between 7.2 to 7.9 [12]. These environmental factors, including humidity, rainfall, temperature, tidal levels, and water pH, collectively contribute to the unique and dynamic ecosystem of the mangrove forest. Therefore, the study is focused on the coastline that is affected.

3. Material and method

The present study uses a collection 2 level 1 images from LANDSAT series of satellites available from USGS Earth Explorer website from 1975 to 2022. To get cloud-free clear images of the study area, images are quired in the month of December each year where possible, or in adjacent months. For the years 1975 and 1980, images were selected from the Landsat Multi-Spectral Scanner (MSS), and the rest images were from Landsat Thematic Mapper (TM, 4 and 5) and Enhanced TM (ETM+) (**Figure 1**). Two adjacent Landsat images of path 148 rows 45 and 147 rows 45 for Landsat MSS; path 138 rows 45 and 137 rows 45 for TM and ETM+ are needed for the Sundarbans coastline of Bangladesh and India (**Table 1**).

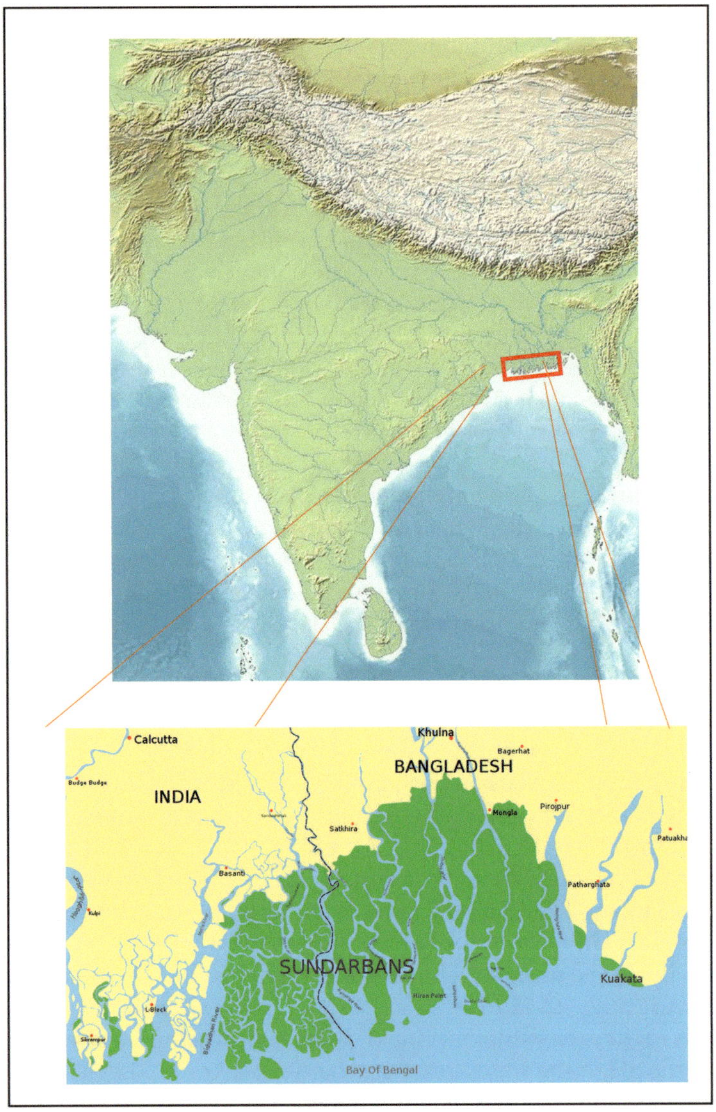

Figure 1.
Location of the study area: Sundarban coastline along bay of Bengal.

Year	Month	Satellite	Path	Cloud cover	Resolution (M)
1975	December	Landsat MSS 1	148/45147/45	0	60
1988	December	Landsat MSS 4	148/45147/45	0	30
1994	December	Landsat MSS 4	148/45147/45	0	30
2005	December	Landsat 7 ETM+	148/45147/45	<7%	30
2015	December	Landsat 8 OLI	148/45147/45	0	30
2022	December	Landsat 8 OLI	148/45147/45	0	30

Table 1.
Landsat images used in the study.

Images downloaded from USGS Earth Explorer are first georeferenced with a final georeferenced image having <±0.5 pixel root mean squared error (RMSE). As the adjacent images fall in two different countries with different UTM Zones of 46 N and 45 N The images are reprojected to Lambert Azimuthal Equal area projection to preserve the area of individual polygon and a true sense of direction from the center. This projection is also preferred for statistical analysis of land change. To maintain the spectral integrity of the image nearest neighbor resampling was used. All images from Landsat MSS1 are resampled to 30 m resolution. Resampling the 60 m MSS pixels to 30 m does not impact the spatial resolution of the images, whereas resampling the 30 m TM pixels to 60 m MSS pixels would degrade spatial resolution of the images.

To enable accurate classification and change detection from multi-temporal satellite imagery, it is crucial to perform radiometric calibration. This process involves correcting for gain and bias variations in the satellite data. In the case of Landsat data, the scattering effect is particularly prominent [13]. Additionally, for vegetation cover identification, atmospheric correction is necessary to mitigate the impact of scattering.

In our study, we conducted atmospheric correction on the Landsat visible and near-infrared (VNIR) bands. This correction involved radiometric calibration, which transformed the digital number (DN) values of the bands into the top of the atmosphere radiance (LTOA) using a sensor calibration function (Eq. 1) proposed by Chander et al. [14]. Subsequently, we converted the radiance of the VNIR bands into accurate surface reflectance using an image-based atmospheric correction model developed by Chavez [15]. This model was chosen for its simplicity and because radiosounding data was not readily available (Eq. 2).

$$L_{TOA} = \left(\frac{L\max_{\lambda} - L\min_{\lambda}}{QCAL\max - QCAL\min} \right) \times \left(DN - QCAL\min \right) + L\min_{\lambda} \tag{1}$$

Where $L\max_{\lambda}$ and $L\min_{\lambda}$ represent the maximum and minimum radiance (in $W/m^{-2}\,sr^{-1}\,\mu m^{-1}$), QCALmax and QCALmin represent the maximum and minimum DN value possible (255/1).

$$\rho = \frac{\left(L_{TOA} - L_p \right) \pi d^2}{ESUN_{\lambda} \cos\theta_z T_z} \tag{2}$$

Where ρ represents the surface reflectance. d denotes the Earth-sun distance, which is measured in Astronomical Units (AU). ESUNλ refers to the band-pass solar irradiance at the top of the atmosphere (TOA) for a specific wavelength (λ); Z represents the solar zenith angle, measured in degrees; TZ represents the atmospheric transmission between the ground and the TOA. For band 4, the value of TZ is assumed as 0.85, while for band 5, it is taken as 0.95 (**Figure 2**) [15]. Lp represents the radiance that results from the interaction of aerosols and atmospheric particles. Its estimation is based on the studies conducted by Song et al. [13], Chavez [15], and Sobrino et al. [16].

FCC (false-color composite) images are commonly used in remote sensing and satellite imaging to enhance the interpretation of land cover and vegetation.

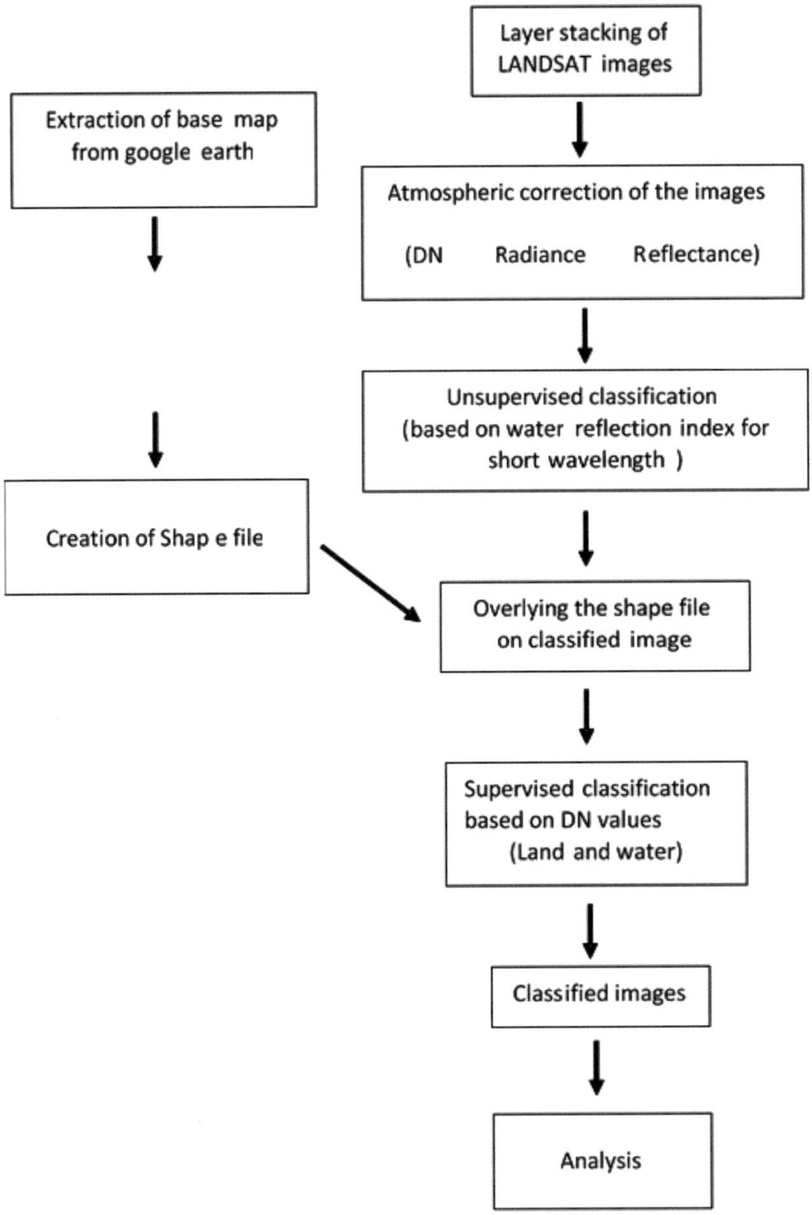

Figure 2.
Overall methodology of the study.

These images are created by combining different bands of electromagnetic radiation, typically in the green, red, and near-infrared regions of the spectrum. The interpretation of FCC images relies on the fact that different materials reflect and absorb different wavelengths of light. In the case of distinguishing between land and water, the choice of bands is important. By using the green, red, and near-infrared bands, it becomes possible to differentiate between various land cover types. In an FCC image, vegetated areas appear in shades of red. This is because healthy vegetation strongly reflects near-infrared light while absorbing more of the red light. As a result, when the near-infrared band is assigned to the red channel in the composite image, it gives vegetation a distinct red color. Bare soils, on the other hand, appear in tones of brown. Since bare soils have little vegetation cover, they reflect both green and red light, giving them a brownish appearance in the composite image. Mudflats or sandy beaches generally appear as shades of white in an FCC image. These areas have a high reflectance in both the green and red bands, resulting in a bright appearance. Water bodies, such as lakes or oceans, appear blue or black in the FCC image. This is because water absorbs near-infrared light, and in the absence of strong vegetation reflectance, it reflects more of the blue and green light. Therefore, water bodies tend to appear darker compared to other land cover types. In addition to FCC images, the normalized difference vegetation index (NDVI) can be calculated using the red and near-infrared bands. The NDVI is a quantitative measure of vegetation health and density. It is computed by taking the difference between the near-infrared and red reflectance values and dividing it by their sum. The resulting NDVI values can help confirm the land-water boundary. Water bodies typically have a negative NDVI value, indicating the absence of vegetation, while dry land usually has a positive NDVI value, reflecting the presence of vegetation.

Overall, FCC images and the NDVI provide valuable information for land cover classification, vegetation monitoring, and understanding the distribution of different land and water features in remote sensing applications.

4. Error estimation

The two main independent sources of uncertainty mentioned in your statement are the uncertainty caused by the georeferencing process and the uncertainty caused by the digitizing process. These two independent sources of uncertainty, namely georeferencing and digitizing, contribute to the overall uncertainty in the spatial data analysis and should be considered when interpreting or using the data. The georeferencing process involves aligning spatial data to a known coordinate system or reference imagery. The error associated with this process is assumed to be normally distributed with a mean of 0 and a standard deviation equal to the Root Mean Square Error (RMSE) resulting from the georeferencing procedure. In this case, the RMSE is ±0.5 pixels. The assumption of normal distribution implies that the errors are symmetrically distributed around the mean, and the RMSE provides an estimate of the typical magnitude of the georeferencing errors. Digitizing refers to the process of converting analogue or physical data into digital form. In this context, it involves delineating the boundary of an area on a map or image. The error associated with this process is assumed to be uniformly distributed, ranging between 0 square meters and 900 square meters. This assumption implies that the digitizing errors have an equal likelihood of occurring within this range. A mixed pixel occurs when a pixel represents a mixture of different land cover types, in this case, soil and water along the coastline. This can be a result of the spatial resolution of the data, where a single pixel covers an area that includes both land and water. The presence of mixed pixels along the coastline can introduce additional uncertainty in the analysis, as the classification or interpretation of such pixels becomes more challenging (**Figure 3**).

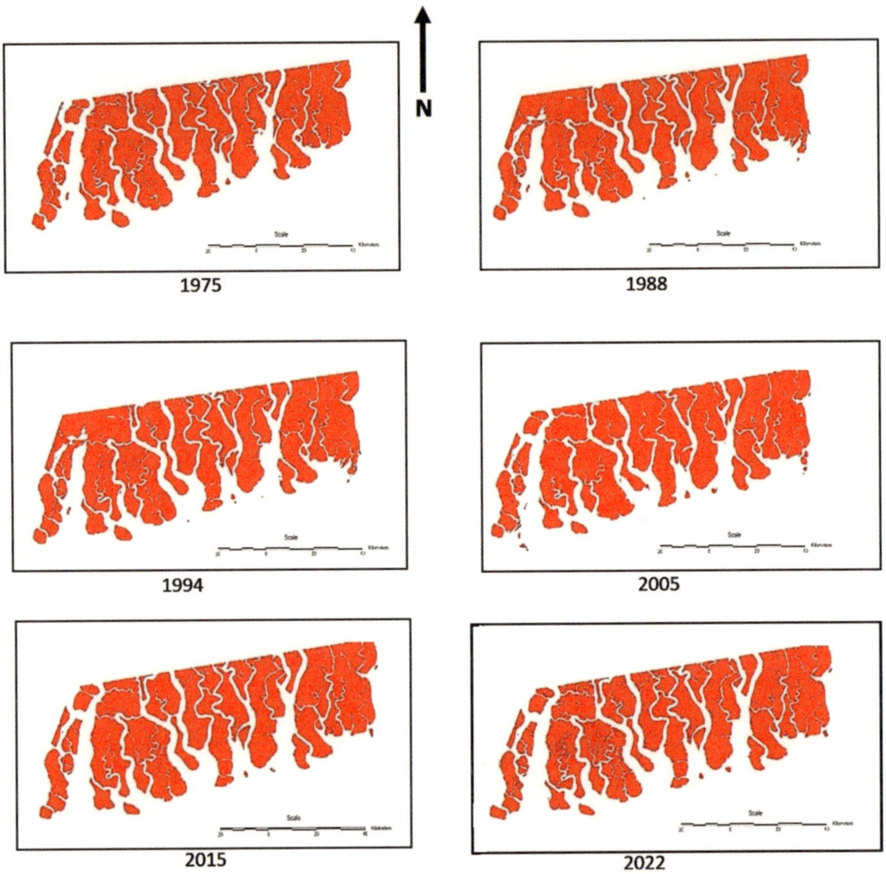

Figure 3.
Changes in coastline from 1975 to 2022.

5. Results and discussion

Accretion and erosion were not the same throughout the different segments of time considered for the study. For example, erosion was dominating in the 1975–1988 interval with a land loss of ~12 km^2/year, while erosion declined in the following periods to a rate of ~8 km^2 per year. If we look at results as a whole erosion wins over accretion and the coast registers a net land loss of ~280 km^2 in this study period. Accretions in the 1994–2005 and 2005–2015 periods were ~ 10 km^2 and ~ 5 km^2 per year respectively. In this segment, a total gain of ~210 km^2 of land can be observed.

According to the present study, the erosion and accretion rates in the Subarnarekha Delta have been highly dynamic over the past 47 years. It is observed that the delta has experienced both erosion and accretion at different times during this period. The total land change estimated over the last 47 years in the Subarnarekha Delta is a loss of 250 km2 of land. This suggests that the delta experienced a net loss of land during this time period. The study also indicates that accretion has been dominant in the east direction, while erosion has been more pronounced in the south-to-west direction. This spatial variation in erosion and accretion patterns suggests that different parts of the delta have been subjected to varying degrees of land loss or gain. These findings highlight the dynamic nature of the Subarnarekha Delta, with ongoing processes of erosion and accretion shaping its coastline over the years. Understanding the spatial patterns and rates of erosion and accretion is crucial for managing and mitigating the impacts of coastal changes in the delta region.

A notable pattern observed in the land dynamics of the Sundarbans coastline is the declining rate of accretion in successive periods. This trend could be attributed to the overall sediment deprivation of the delta caused by human activities such as dam construction and other anthropogenic disturbances upstream (**Table 2**) [17].

Erosion is observed in all directions except for the landward directions (N, NE, NW), suggesting the potential influence of sea level rise (SLR) impacts in the Bay of Bengal. The variation in erosion rates in different azimuthal directions may be attributed to a combination of surface wave and tidal actions. Surface waves primarily originate from the southwest direction in the Bay of Bengal, while tidal actions predominantly occur from the south [18]. Additionally, the East India Coastal Current (EICC) flows northward during the rainy seasons (March–September) and reverses its direction during the dry seasons (October–January). As a result, the combined forces of waves and tides are stronger in the south azimuthal direction compared to other directions.

These factors contribute to the complex interplay of erosion and accretion along the Sundarbans coastline, with variations in sediment dynamics influenced by SLR, wave action, tidal forces, and seasonal currents (**Figures 4** and **5**).

Since there is no dike or other construction to safeguard the shoreline, the coastline retreat in the Sundarban area was expected due to sea level rise. By storing sediments upland and reducing their availability at the shore, sediments from dam construction have also had a substantial impact. While tidal action occurs from the south, surface waves in Sundarbans coastline region are primarily from the southwest. The East India Coastal Current (EICC) flows southward during the dry seasons of October to January and northward during the rainy seasons of March to September. The southern region is more affected by the combined impact of waves and tides than other regions. During the final 47 years of the research period, the Subarnarekha coast lost 270 km2 in total, or 5.7 km2 per year. This number is higher than what preceding researchers had predicted. There has been a loss of land, but not equally. Despite the fact that some new islands have appeared along the coast, the overall picture shows a loss of land mass there.

Period	Accretion (Km^2 $year^{-1}$)	Total accretion (km^2)	Erosion (km^2 $year^{-1}$)	Total erosion (km^2)	Difference (km^2)
1975–1988	7.3	92.7 ± 0.8	12.9	167.8 ± 0.8	−75.1
1988–1994	6.58	32.9 ± 0.5	10.94	54.7 ± 0.3	−21.8
1994–2005	10.22	112.5 ± 0.4	17.27	190.2 ± 0.3	−77.7
2005–2015	6.1	61.5 ± 1.3	11.5	115.5 ± 1.6	−54
2015–2022	4.18	29.3 ± 0.5	8.51	59.6 ± 0.9	−30.3

Table 2.
Accretion and erosion rates in the Sundarbans coastline estimated.

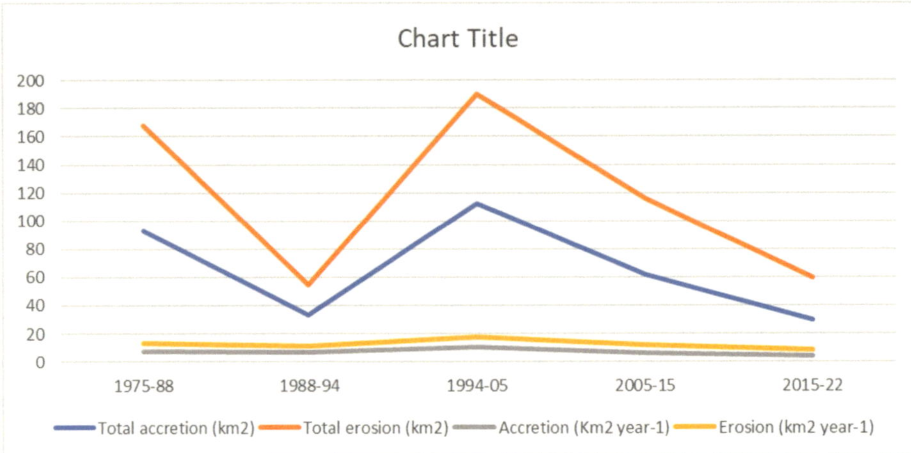

Figure 4.
Erosion and accretion over the past 47 years in suburban coastal area.

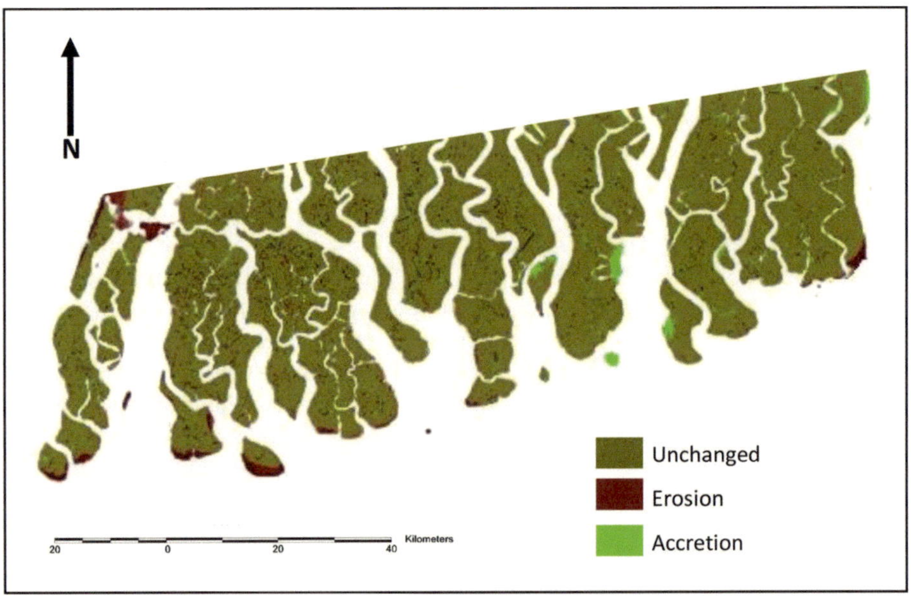

Figure 5.
Accretion and erosion of the coastline between 1975 and 2022.

6. Conclusions

Our research emphasizes the intricate nature of the spatiotemporal dynamics observed along the retreating Sundarbans coastline in the Bengal delta. To delineate the coastline, we utilized cloud-free Landsat images and developed algorithms that consistently derived distances and areas of land dynamics for the entire coastline at regular intervals. While coastal retreat is a natural global phenomenon associated with sea-level rise (SLR), our study delved into the specific impacts of SLR, as well as reduced discharge and sediment flow from the contributing river, on the coastline.

DOI: http://dx.doi.org/10.5772/intechopen.112243

The formation of the Bengal delta was primarily driven by the discharge of the Ganges-Brahmaputra (GB) river, resulting in accretion dominating the region for thousands of years. However, while some previous sampling studies suggested ongoing accretion in the Bengal delta [11], recent modeling studies indicate that sediment compaction has caused the delta to sink [17]. Our study presents the first evidence that the entire non-diked portion of the Sundarbans coastline in the Bengal delta is currently experiencing a net erosional state.

By analyzing a time series of satellite images, we were able to characterize the spatial and temporal aspects of the retreat. This approach reduced uncertainties inherent in modeling and sampling studies of continuous spatial processes such as coastal dynamics. The spatiotemporal analysis conducted in our study may facilitate future research in understanding the local and global factors contributing to the reported spatial variations in erosion and accretion.

We anticipate that the findings of our study will have practical implications for the management planning of the Sundarbans—the world's largest remaining patch of mangrove forests.

Author details

Partha Sarathi Mahato
Department of Geology, Government General Degree College Gopiballavpur-II, Jhargram, West Bengal, India

*Address all correspondence to: parthamahato1@gmail.com

References

[1] Kuenzer C, Bluemel A, Gebhardt S, Quoc TV, Dech S. Remote sensing of mangrove ecosystems: A review. Remote Sensing. 2011;**3**:878-928

[2] Upadhyay VP, Ranjan R, Shing JS. Human mangrove conflicts: The way out. Current Science. 2002;**83**:1328-1336

[3] Blasco F, Aizpuru M. Classification and evolution of the mangroves of India. Tropical Ecology. 1997;**38**:357-374

[4] Naskar KR. Manual of Indian Mangroves. New Delhi: Daya Publishing House; 2004

[5] Eriscson JP, Vörösmarty CJ, Dingman SL, Ward LG, Meybeck M. Effective Sea-level rise and deltas: Causes of change and human dimension implications. Global and Planetary Change. 2006;**50**(1-2):63-82

[6] Nandy S, Kushwaha SPS. Study on the utility of IRS 1D LISS-III data and the classification techniques for mapping of Sundarban mangroves. Journal of Coastal Conservation. 2011;**15**:123-137

[7] Green EP, Clark CD, Mumby PJ, Edwards AJ, Ellis AC. Remote sensing techniques for mangrove mapping. International Journal of Remote Sensing. 1998;**19**:935-956

[8] Dwivedi RS, Rao BRM, Bhattacharya S. Mapping wetlands of the Sundarban delta and its environs using ERS-1 SAR data. International Journal of Remote Sensing. 1999;**20**:2235-2247

[9] Blasco F, Aizpuru M, Gers C. Depletion of the mangroves of continental Asia. Wetlands Ecology and Management. 2001;**9**:245-256

[10] Allison MA. Historical changes in the Ganges–Brahmaputra delta front. Journal of Coastal Research. 1998;**14**(4):1269-1275

[11] Allison MA, Kepple EB. Modern sediment supply to the lower delta plain of the Ganges–Brahmaputra River in Bangladesh. Geo-Marine Letters. 2001;**21**:66-74

[12] Banerjee LK. In: Shing NP, Shing KP, editors. Floristic Diversity and Conservation Strategies in India. Vol. V. Botanical survey of India. Ministry of Environment and Forests; 2002. pp. 2801-2829

[13] Song C, Woodcock CE, Seto KC, Lenney MP, Macombe SA. Classification and change detection using Landsat TM data: When and how to correct atmospheric effects? Remote Sensing of Environment. 2001;**5**:230-244

[14] Chander G, Markham BL, Barsi JA. Revised Landsat- 5 thematic mapper radiometric calibration. IEEE Geoscience and Remote Sensing Letters. 2007;**4**:490-494

[15] Chavez PS Jr. Image-based atmospheric corrections – Revisited and improved. Photogrammetric Engineering and Remote Sensing. 1996;**62**(9):1025-1036

[16] Sobrino JA, Jimenez-Munoz JC, Paolini L. Land surface temperature retrieval from LANDSAT TM 5. Remote Sensing of Environment. 2004;**90**:434-440

[17] Syvitski JPM, Kettner AJ, Overeem I, Hutton EWH, Hannon MT, et al. Sinking deltas due to human activities. Nature Geoscience. 2009;**2**:681-686. DOI: 10.1038/ngeo629

[18] Poterma JT, Luther ME, O'Brien JJ. The seasonal circulation of the upper ocean in the bay of Bengal. Journal of Geophysical Research. 1999a;**96**(C7): 12,667-12,683

Chapter 8

Drivers and New Opportunities for Woody Vegetation Use in Erosion Management in Pastoral Hill Country in New Zealand

Ian McIvor, Thomas Mackay-Smith and Raphael Spiekermann

Abstract

Increases in the magnitude and frequency of rainfall events in New Zealand due to climate change, coupled with existing concerns about sediment and nutrient contamination of waterways, are changing policy and practice around erosion management and land use. We describe the challenges around slope erosion reduction, cover current legislation and management practices, illustrate how modeling can inform erosion management and describe new opportunities, whereby native species can become a new active management tool for erosion control. Passive erosion management depending on natural revegetation by slow growing woody species is used on land retired from grazing but is much less effective than active erosion management in reducing shallow slope erosion. Active erosion management using exotic fast-growing poplar and willow trees strategically placed on hillslopes is effective in reducing erosion, but these trees can be hard to establish on drier upper slopes. An endemic woody tree, Kanuka, grows on drier slopes and is being tested as an erosion control tool. Kanuka seedlings have been successfully established on pastoral slopes, including drier slopes. A spatial decision support tool developed to identify pastoral hillslopes at high risk of erosion has improved decision-making when positioning appropriate trees on these slopes.

Keywords: erosion management, landslide susceptibility, poplar, native vegetation, kānuka

1. Introduction

Following arrival by European settlers in the 1800s, extensive deforestation for pastoral farming resulted in a geomorphic landscape response consisting of high erosion and sedimentation rates [1–5]. Increased erosion rates have led to a variety of adverse consequences, including i) reduced land productivity, ecosystem services, and food security through loss of productive soil; ii) increased damage to infrastructure; iii) adverse impacts on water quality and aquatic ecosystems from increased sediment delivery to streams; and iv) negative impacts for cultural values related to soil and aquatic environments [6–13]. Erosion processes in New Zealand remain very active [14], due to a predisposed natural environment with steep slopes, weak sedimentary rocks, and a climate featuring high annual rainfall and relatively frequent high magnitude rainfall events [15–17]. Climate change is predicted to result in large increases in sediment loads, primarily due to increasing storm magnitude-frequency of mass movement erosion in soft-rock hill country [18].

Mass movement processes are geographically the most widespread type of erosion in New Zealand [15]. The most common types of mass movement in New Zealand are shallow, rapid slides and flows involving soil and regolith [1, 19]. Such landslides are generally triggered either by high-intensity-rainfall events or by small rainfall events on top of saturated soil moisture conditions [18]. Shallow landslides make up the largest source of sediment from pastoral hill country in New Zealand [20]. In these steep and highly dissected pastoral landscapes, bioengineering—either through widely spaced trees, blanket afforestation, or through natural reversion to indigenous forest—has been the most common method to increase slope stability and reduce soil erosion [21–23]. Tree roots are more effective than pasture roots in binding soil and preventing shallow landslides in pastoral landscapes [21].

Sheep and beef production depends on hill country for the supply of breeding stock as well as prime animals for meat processing. Wool has been a significant commodity product of hill country farms in the past and may well regain importance in future. Converting native forest to farmland was seen as a necessary activity to provide livelihoods for new immigrants and generate national wealth from the supply of essential export products, wool, and meat. The conversion happened quickly without an awareness of the inherent instability of the slopes and their vulnerability to landslides when the soil becomes saturated.

Hill country landscapes in scope for this chapter include those low altitude lands (<1000 m a.s.l.) that feature rolling and steep slopes (>15°), are not regularly cultivated on a large scale, are dominated by diverse pasture systems (but may include various woody vegetation components), and are managed for mixed livestock operations (mainly sheep, cattle, and deer) [24]. The area covered by this loose definition is about 5.2 million ha [25] or approximately 20% of New Zealand. Most of these landscapes have been developed into productive pastures from indigenous broadleaf-podocarp forest over the last century, but in many cases, the prevailing vegetation has seen cycles of reversion to scrub, or establishment of plantation forestry as the economic and social drivers have shifted over decadal scales [23]. While active measures to reduce soil erosion in pastoral hill country have been undertaken at central and local government level, the land remains in private ownership, and as such, erosion management is dependent on individual landowners carrying out erosion control measures to stabilize their pastoral slopes.

The key contaminants for hill land waterways are sediment, P, N, and fecal microorganisms. Sediment loss from large-scale erosion events, in terms of both the immediate and ongoing quantities of soil loss, is the biggest environmental management issue for hill country [23]. Phosphorus is included ahead of N as most surface waters in New Zealand are more P-limited than N-limited [26], and total P losses in hill environments are strongly linked to sediment [27]. In general, relative to waterways in forested catchments waterways draining pastoral-dominant catchments have greater water yields, peak flows, nutrient concentrations, suspended sediment concentrations, and fecal coliform concentrations [28].

2. Drivers for erosion management in pastoral hill country

2.1 Water quality

In common with the rest of the world, New Zealand rates water quality of its natural waterways with high priority. Both P and N are considered contaminants of natural waterways, and enrichment does generate eutrophication conditions particularly during periods of low flow. Likewise, sediment sourced from land is detrimental to aquatic life, reducing clarity and contributing P.

2.2 Soil loss

Soil fertility is largely held within the topsoil. Recovery of soil fertility once topsoil is removed from pastoral slopes in shallow landslides is a very slow process [29] and has serious consequences for soil health, pasture growth, carbon storage, and rural livelihoods.

2.3 Asset protection

Shallow landslides can be very damaging to pasture, farm tracks, drainage channels, fences, and possibly buildings. Stock losses may also occur, though the financial costs are much higher for infrastructure. Public assets such as roadways, bridges, and communication infrastructure are also threatened by severe slope erosion. Our warming climate has increased the risk of tropical weather systems reaching New Zealand. Cyclone Gabrielle devastated parts of northern and eastern North Island in February 2023, transferring large volumes of silt and water from hills to valley floors, destroying homes and livelihoods, and severely disrupting infrastructure. It was preceded by a tropical rainstorm just two weeks prior and saturated soils from an unusually wet spring and summer.

2.4 Carbon credits

New Zealand operates an Emission Trading Scheme (ETS), whereby landowners can gain tradeable carbon credits by planting trees in their landscape to sequester carbon. The requirements (30% canopy cover, tree species height at maturity >5 m, treed area at least 1 hectare, mean width at least 30 m across) (https://www.mpi.govt.nz/forestry/forestry-in-the-emissions-trading-scheme/) provide sufficient tree densities to both gain credits and be effective in preventing shallow landslides.

2.5 Societal sensibility

There is a strong social stigma associated with any action or inaction that damages the natural environment. Rural landowners are held responsible for environmental damage from soil erosion and waterway contamination, and this is a significant driver for erosion management. Many community groups supported by government bodies are planting public areas such as streambanks with native woody vegetation, and rural landowners are being challenged to demonstrate the same environmental awareness.

3. Current approaches to erosion management in pastoral hill country

3.1 Government legislation requirements for erosion management

The Water and Soil Conservation act of 1941 mandated statutory bodies to manage erosion control and flood management. However, under this Act, landowner response to erosion control was largely voluntary.

The Resource Management Act 1991 (RMA) is now the main piece of legislation that sets out how New Zealand should manage its environment. The RMA is based on the idea of the sustainable management of resources and encourages communities and individuals to actively engage in environmental protection.

More recently, the 2020 National Policy Statement for Freshwater Management (NPS-FM) has required regional authorities to manage freshwater in a way that considers the effects of land use, including the effects on receiving estuarine environments (New Zealand Government, 2020). Moreover, Freshwater Farm Plans (FFP) have been established as a legal instrument under the RMA to identify environmental actions on farms in consideration of objectives for the catchment. The Act specifies that an FFP must *"identify any adverse effects of activities carried out on the farm on freshwater and freshwater ecosystems"* and *"specify requirements that (i) are appropriate for the purpose of avoiding, remedying, or mitigating the adverse effects of those activities on freshwater and freshwater ecosystems; and (ii) are clear and measurable"* (Section 217F, RMA). Therefore, understanding the impact of erosion and sediment control is important to achieve the desired environmental outcomes and—more specifically—sediment standards."

The Resource Management Act 1991 and National Policy Statement for Freshwater Management 2020 both require and assist landowners to carry out erosion control and freshwater quality measures on their properties using farm plans and on farm advice and through the provision of woody vegetation planting materials (poplar and willow poles, native plant seedlings) and financial assistance for plant protection.

3.2 Passive erosion management

Financial incentives (e.g., fencing) may be provided to retire steep land with low productivity and high erosion vulnerability [30] from active grazing and allow natural regeneration of woody native vegetation. Native forested catchments have been shown to generate lower soil erosion loads to rivers compared to pasture catchments [31–33]. Quinn and Stroud [33] reported suspended sediment loads of 988 kg ha^{-1} $yr.^{-1}$ in a pasture catchment and 320 kg ha^{-1} $yr.^{-1}$ in a native forest catchment. Furthermore, when compared suspended sediment in a pastoral catchment (180 ha) and native forested catchment (10 ha) on similar topographies and soil types ~8 km apart, Bargh [31, 32] measured loads of 1400 kg ha^{-1} $yr.^{-1}$ in the pastoral catchment and 120 kg ha^{-1} $yr.^{-1}$ in the forested catchment.

Native vegetation regeneration is likely to be a slow, incremental process and of itself leaves the slope no less vulnerable to erosion until any establishing woody vegetation with appropriate root systems can stabilize the slope. In a study on Oashore Station (farm property), Banks Peninsula, 3.3% of the retired area showed observable increase in natural revegetation between 2003 and 2016 despite the property being managed to support natural regeneration [34]. Significant factors influencing natural regeneration were the distance to existing woody vegetation ($p < 0.01$), woody vegetation within 25 m ($p = 0.008$), and years without cattle ($p < 0.01$) with other nonsignificant factors being topographical wetness, years without sheep, solar radiation, slope, elevation, and aspect [34]. These findings are consistent with studies on natural woody regeneration in tropical regions of Brazil where tree cover was found to increase by 0.3–0.4% per year (e.g., [35]).

Passive regeneration, while being slow and incremental, goes some way to satisfying societal aspiration to reduce soil erosion, increase ecosystem biodiversity, improve water quality, and restore parts of the rural landscape to primeval native forest with its incumbent birdlife. For instance, it is estimated that landslides reduced by 65% in hill country with 10-year-old regenerating scrub (mānuka (*Leptospermum scoparium* and kānuka (*Kunzea* spp.)) compared to open pasture, and there was an estimated 90% reduction in landslides for 20–year-old scrub [36]. However, to achieve this protection, scrub density was 20,000 tree ha^{-1} at age 10 years and estimated at 10,000 tree ha^{-1} at age 20 years [36], reducing grazing pasture considerably. In the meantime, the pastoral slope may be invaded by woody exotic species such as barberry (*Berberis* spp.) or gorse (*Ulex europaeus*), which colonize much more readily but which may improve conditions for establishment of native species. Because of small size and slow growth of the woody vegetation, passive natural regeneration is unlikely to generate income from carbon credits under the ETS.

Research is needed across all climatic zones on rates of passive regeneration, and ways to blend passive and active approaches in erosion management.

3.3 Active erosion management

The objective of active erosion management is to target potential sediment source areas with tree planting to increase slope stability and thereby prevent landslide initiation and deposition of debris into adjacent streams. Active erosion management operates at a policy level (e.g., local authorities offering financial assistance to counter erosion, particularly slope erosion, and promoting/producing environmental farm plans as a tool to guide landowners in identifying areas where management activities can effectively reduce erosion) and at an operational level (e.g., providing advice, providing woody plant materials, placing and planting woody plant material, assessing success of planting projects, forming catchment groups of landowners).

3.3.1 How modeling landslide susceptibility informs active erosion management

The policy and operational level represent different scales at which erosion mitigation is planned and implemented. At both levels, the aim is to design and implement cost-effective, targeted erosion control measures to conserve soil and meet water quality targets. Spatial modeling can help make the connection between catchment erosion sources, sediment loads in rivers, and sediment-related water quality. In particular, statistical landslide susceptibility models based on empirical observations of previous landslides (landslide inventories) can help determine where landsliding can be expected in future heavy rainfall events [37, 38]. Statistical models assume that locations with similar physical characteristics to where past failures have occurred are also likely to fail in future. This assumption is tested through validation of models by training in one area and testing in another or by splitting the landslide inventory into train and test sets. By coupling landslide susceptibility models (probability of future landslide occurrence) and sediment connectivity models (probability of sediment-delivery to adjacent streams), potential source areas of landslide-derived sediment can be identified and targeted [39]. The importance of targeted approaches to erosion control was demonstrated by Spiekermann et al. [39] who found a 10-fold increase in cost-effectiveness of targeted mitigation of landslide-derived sediment compared to a non-targeted (random) approach.

Scale considerations are important at different levels of erosion management (policy and operational). Models that use national data inputs are of greater utility at catchment to regional scales in determining where the problem areas are within a particular catchment (**Figure 1**) [37]. Increased spatial refinement of models (e.g., including data on individual trees) can support planning at farm to slope scale by identifying specific locations on a farm that are prone to landslide erosion and/or potential sources of sediment [38].

Following careful erosion mitigation planning, the key activity is planting of woody vegetation within the pastured area, usually on slopes where erosion events are expected to occur in future. Other operational activities exist to support the key activity. Choice of planting material is determined by its compatibility with future use of the land, whether continued pastoral farming or retirement.

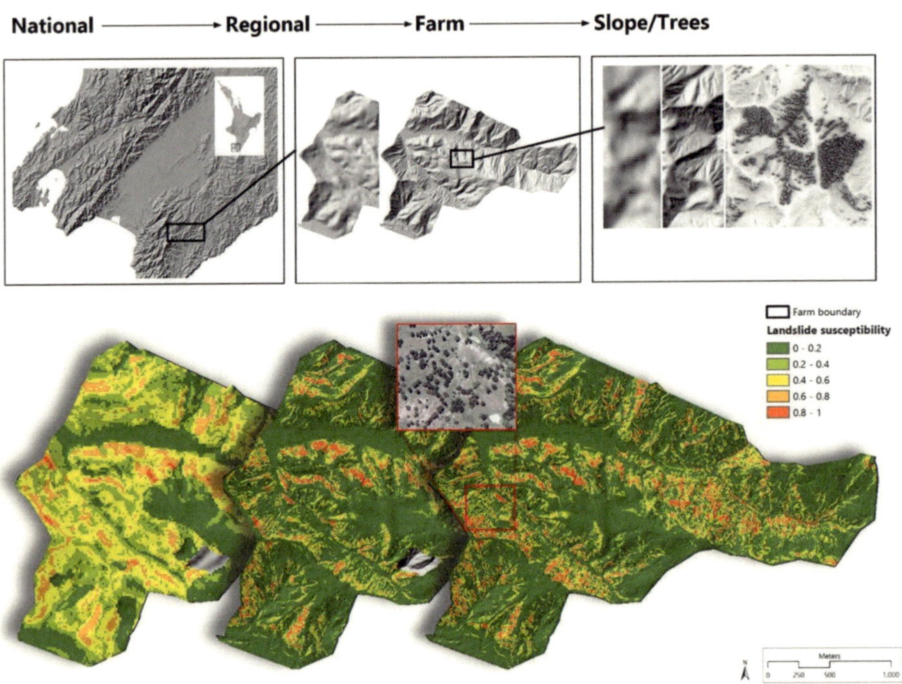

Figure 1.
Scale considerations: Different data products available at different scales can serve different purposes to support decision-making for erosion and sediment mitigation. Shown above are shaded relief maps of digital elevation models (DEMs) at different scales: From left to right: 15-m (based on topographic contour lines), LiDAR-based 5-m and 1-m ground-sampling-distance (GSD). Below are landslide susceptibility models using topographic variables derived from the differently scaled DEMs (15 m, 5 m, 1 m) shown above. The 15 m and 5 m DEM-based models using the land cover database of New Zealand (LCDB) are based on Smith et al. [37], the 1 m model using individual trees based on Spiekermann et al. [38]. Other issues determined by the scale of data used for modeling include its impact on the accuracy/performance of models, as well as requirements of computing power.

3.3.1.1 Spatial decision support for targeted erosion mitigation

3.3.2 Spaced planting of poplar and willow

Poplars and willows have been the species of choice within pastoral systems since the early days of active erosion management. They are planted as 3 m vegetative stem poles in winter when soil moisture is high, and at distances of 8–15 m apart, with the closer spacings applied particularly in gullies with high erosion risk. Each pole is planted in such a way as to capture surface runoff, avoid stock paths, and promote root development (https://www.horizons.govt.nz/HRC/media/Media/Land/Growing-poplars-info-sheet-2014.pdf). The pole is protected from any browsing activity by sheep but vulnerable to browsing by larger animals. For this reason, it is recommended that cattle be excluded from planted areas for two years. Survival rates for poles are >90% in good years, and annual height growth from poles can be 2–3 m (**Figures 2** and **3**). The practice has limitations in that higher positions on the slope are more difficult for poplars to establish because of shallow soil depth, low soil moisture, greater exposure to wind, or little topsoil resulting from a previous erosion event.

Mature poplars and willows planted at low densities (<70 tree ha^{-1}) reduced landslide occurrence in pastoral hill country by 95% [21] compared with pasture-only protection.

3.3.3 Change of land use to commercial forestry

Whole farm conversion from traditional pastoral farming in hill country to either production forestry or carbon forestry has accelerated in New Zealand since 2011 (https://www.rnz.co.nz/news/country/473898/overseas-firms-buy-more-sheep-beef-farms-for-forestry-conversion), prompting a change in regulations for foreign investors. The primary forestry species that is planted is radiata pine (*Pinus radiata*).

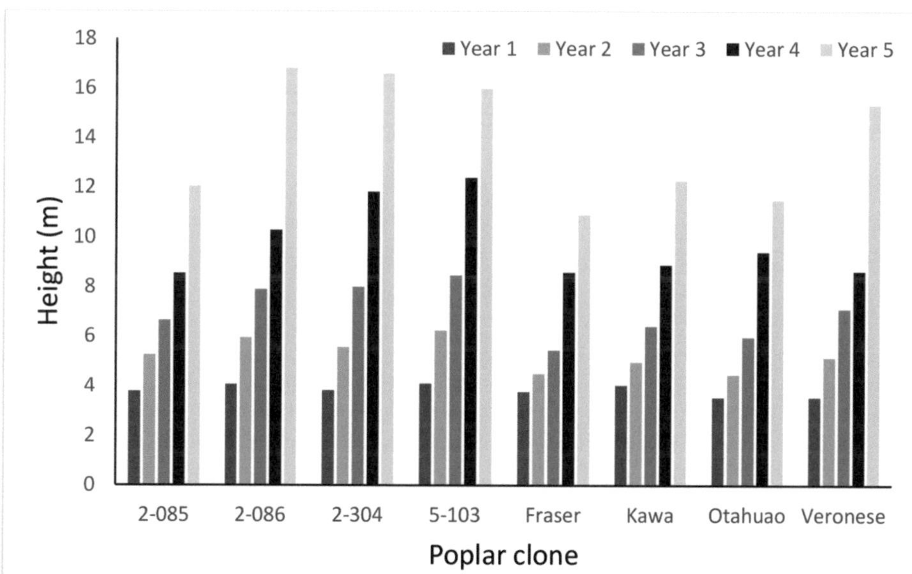

Figure 2.
Mean height (m) in the first five years of poplar clones growing on a pastoral hill slope.

Figure 3.
Mixed clone poplars at age five years and planted as 3 m poles.

Carbon forests are expected to be profitable within the carbon market, without the market uncertainties associated with pastoral farming and without the risks of slope erosion associated with harvesting. Since this trend is recent, there is much debate about the future consequences of establishing permanent forests of fast-growing pine trees on valuable pastoral land. Furthermore, there are concerns with the downstream effects of 'slash' (the woody material left behind after the trees are harvested) after storm events, which is deposited in rivers and beaches (https://environment.govt.nz/what-government-is-doing/areas-of-work/land/ministerial-inquiry-into-land-use). Close planted forests have been shown to be more effective than space-planted poplars or willows in reducing slope erosion, moderating runoff, and improving water quality [15]. An alternative approach promoted by New Zealand Farm Forestry Association is to plant slopes vulnerable to erosion with close-planted timber tree species while retaining gentler slopes and less erosion-prone land for pastoral farming.

3.3.4 Planting native vegetation

Another option that is used less frequently than commercial forestry is planting high density (> 1000 tree ha^{-1}) stands of native vegetation. Instead of natural regeneration, mānuka is intentionally planted, providing an alternative opportunity for income from either honey or essential oil production [40]. Climax forest species can also be planted, but this is expensive, costing ~NZD$10,000 per ha compared to ~NZD$1650 per ha for establishing radiata pine [41]. A cheaper alternative to both these options is letting the land regenerate passively, as discussed in Section 3.2.

4. A novel approach of space planting with native woody vegetation

Native vegetation already grows on many parts of grazed hill country [42], having established naturally via seeds, and provides a valuable slope stabilization role (**Figure 4**). Native trees are typically not planted in space-planted systems in hill country. One likely reason for native trees not being regularly planted in space-planted systems is the difficulty of planting and protecting native seedlings in the presence of livestock. Despite this, there has been a growing interest in using natives in hill country silvopastoral systems [43, 44], although there still remains a paucity of methodology on establishing native trees in space-planted systems in hill country. Considering this, an establishment trial was undertaken to learn whether it is possible to protect and grow kānuka seedlings in the presence of livestock, with kānuka being one of the most common natives already growing in hill country [42]. This research is ongoing, but this chapter presents preliminary results from this trial.

Figure 4.
Mature kānuka trees growing in hill country (top left), spaced kānuka seedlings on erosion-prone N slope at Gladstone (top right), a kānuka seedling enclosed in its protector (bottom left), and a kānuka seedling a to (bottom right).

4.1 Case study establishing space-planted kanuka on pastoral land

In July 2021, 30 kānuka seedlings (*Kunzea ericoides*) were planted on two aspects (North and South) on three hill country pastoral farms with contrasting rainfall (**Table 1**). Seedlings were grown in root trainer pots and at planting were on average 52 cm tall (minimum: 29 cm; maximum: 69 cm). The seedlings were protected using a plastic mesh tube and supported by two steel Y-posts and a steel rebar (**Figure 4**). Weather stations were installed on both aspects at each farm to measure rainfall. The weather station rain sensors at the Gladstone sites malfunctioned on both aspects, so rainfall data for this farm was used from a local weather station. Livestock (sheep, cattle) were continuously grazing at all sites. There have been three measurements so far on the trees (**Table 1**), although the third measurement (t3) was not undertaken at Ahuriri because a recent storm event (Cyclone Gabrielle) had made the site inaccessible. Any seedling deaths between the planting date and measurement 1 (t1) were likely related to root desiccation following planting.

Rainfall at the Ahuriri north and south slopes was 1447 and 1417 mm in 2022, respectively, and it was 1807 and 1828 mm at Taumarunui north and south, respectively. The rainfall at the Gladstone site was 1230 mm in 2022. The mean heights at t1, t2, and t3 for all the seedlings over all the sites were 62.7, 93.3, and 153.1 cm, respectively (**Figure 5**). This represented an average growth of 9.9, 40.8, and 100 cm at t1, t2, and t3, respectively. Seedling survival was varied. At t3, 70% of the seedlings survived at Gladstone north, 90% survived at Gladstone south, 96.4% survived at Taumarunui north, and 93.3% at Taumarunui south. At t2, 83.3% of the seedlings survived at Ahuriri north and 72.0% survived at Ahuriri south. There were 0% shock deaths at Taumarunui, 8.3% at Gladstone, and 21.8% at Ahuriri. Livestock damaged the protectors of 0% of the seedlings at Gladstone, 8.3% at Ahuriri, and 4.8% at Taumarunui.

It is likely that the higher rainfall at Taumarunui resulted in the higher growth rate of kānuka at this farm at t3 compared to Gladstone. Kānuka growth was higher on

Site	Location	Elevation (masl)	Average rainfall (mm, 2002–2022)	Planting date	Measure1 (t1)	Measure 2 (t2)	Measure 3 (t3)
Gladstone north	175.74° E, 41.05° S	300	898	27/07/21	04/01/22	11/10/22	21/02/23
Gladstone south	175.74° E, 41.05° S	300	898	27/07/21	04/01/22	11/10/22	21/02/23
Ahuriri north	176.791° E, 39.479° S	48	817	22/07/21	18/12/21	13/09/22	n/a
Ahuriri south	176.779° E, 39.475° S	116	817	22/07/21	18/12/21	13/09/22	n/a
Taumarunui north	175.281° E, 38.904	333	1773	13/07/21	21/12/21	04/07/22	21/03/23
Taumarunui south	175.281 E, 38.897 S	328	1773)	13/07/21	21/12/21	04/07/22	21/03/23

Table 1.
Site information, planting and measurement dates for the kānuka establishment trial.

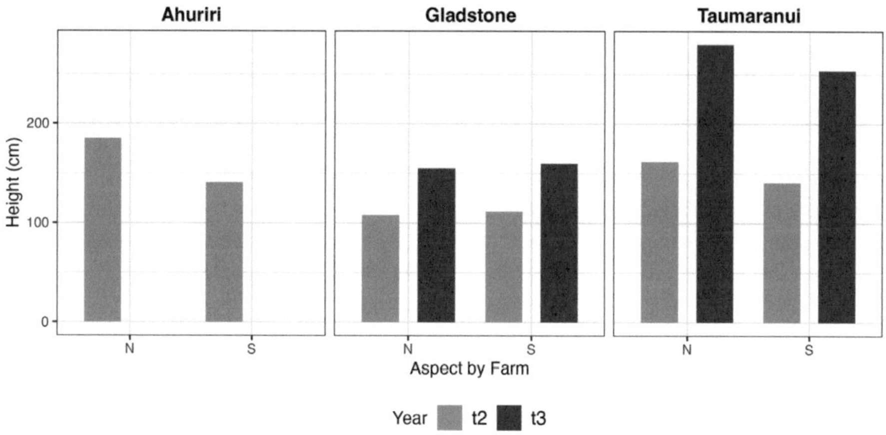

Figure 5.
Kānuka heights after measurements t2 and t3 for each aspect (north and south) and the three farms (Ahuriri, Gladstone, Taumarunui).

the drier and warmer North facing slopes, though the two drought-prone sites had exceptional wet summers in both 2021–2022 and 2022–2023. Further analysis after the trial has finished will compare rainfall data with morphometric data (e.g., slope) to provide a better understanding as to the drivers of seedling survival and growth. The average growth rates for the poplar clones in **Figure 1** after year 2 were between 2.0 and 6.2 m. The kānuka seedlings grew on average 1.5 m at t3, although will likely grow slightly more until the end of year 2 (July 2022). This gives evidence that the kānuka would stabilize slopes more slowly than poplar. However, kānuka may take a similar time to reach maturity height because kānuka is a smaller tree (10–20 m) than poplar (25–35 m) when fully grown. Rate of root extension of young kānuka or root biomass and distribution of mature kanuka in space planted systems, features important for erosion control, are yet to be researched.

5. Pros and cons of different space-planted approaches

There are important reasons why poplar and willow have been adopted as space-planted trees for erosion control in New Zealand. They are highly effective at reducing landslides, generally grow 2–3 m per year, and they can be planted as 3 m unrooted poles in the presence of livestock. The plastic protection sleeves are easy to apply, are durable, and split apart as the tree grows. However, few of the discarded sleeves are recycled, so they are being reviewed as a pollutant and not sustainable. The trees provide a highly palatable drought fodder source, and shed leaves are consumed by stock. Tree shading reduces pasture production considerably as the trees age [45], but this effect was less significant on poorly grassed slopes, and pasture production per hectare reduced by only 7–12%. (https://www.poplarandwillow.org.nz/documents/influence-of-shading-by-poplar-trees-on-pasture-production-rb02.pdf). Removing lower branches to create a timber tree or pollarding for fodder production are management techniques employed to reduce shading and promote pasture production under the trees.

The preliminary results of the kānuka trial show that it is possible to establish native seedlings in hill country; however, survival rates were mixed, and the best current protection method is time-consuming to use. Forming a better understanding as to why survival rates were low on some of the aspects and improving on the current protection method will be essential for the uptake of native silvopastoral systems with kānuka in New Zealand. Also, it is important to extend research to other native species in space-planted systems to assess the viability of native space-planted systems more generally. These considerations are very relevant to the indigenous Maori wanting to develop their pastoral land use to include spaced plantings of native woody species used for traditional medicines. Research initiatives that strengthen Maori culture and livelihoods are encouraged by the New Zealand government.

In terms of other co-benefits of trees, a recent study has reported pasture production to be over 100% greater under isolated and mature kanuka trees compared to open pasture at two hill country sites [44]. Although these results are preliminary and require further validation on other sites, they indicate there could be advantages when using kānuka in space-planted systems in New Zealand when compared to poplar and willow, with research finding that poplars greater than 15 years old negatively impact pasture production between 12 and 65% [45]. It is likely there will always be a place for faster growing space-planted trees such as poplar and willow, especially in areas that are highly susceptible to erosion. However, when they can be established more easily, other native species could provide other benefits to areas that are less susceptible to erosion and that do not require immediate erosion protection. Mature kānuka can be found growing higher up drier north-facing slopes, locations generally too dry for poplars and willows to establish with the current methods. It may be that kānuka seedlings will establish in these locations. Further research will test this hypothesis.

Spatial decision support tools for targeted erosion mitigation are of particular benefit to regional land managers advising landowners on species choice and location of plantings and providing landowners with a reasoned cost/benefit assessment when planting tree in high risk, low survival positions on the slope.

6. Support for landowners to reduce erosion

6.1 Institutional support

New Zealand Government One Billion Trees Fund, due to end in 2028, targeted at tree planting projects to help meet international climate change commitments, among its broader aims, funds trees planted for erosion control on pastoral land. Extensive areas of native woody species are being planted for active erosion management by this fund. The Hill Country Erosion Fund allocated from Central Government and administered by Regional Authorities supports both passive and active erosion management by co-funding planting and retirement initiatives by landowners.

6.2 Technology support

Researchers provide high-quality baseline data (understanding of erosion processes, tree-soil interactions, state of water resources) and develop new approaches to overcome economic and environmental conflicts. Extension support through publications and on-site advice are offered at the local level to encourage active erosion management. Regional Authorities provide planting materials and protectors to best practice specifications, and success is the highest when this extends to on-site placement, planting, and subsequent tree management as happens in some regions.

6.3 Demographic support

Landowners traditionally think in terms of their own properties without considering landscapes or catchments. The formation and funding of Community Catchment Groups is proving effective in reducing this barrier as landowner collectives who identify with a geographical area, usually a local catchment, use the group funds to cooperate to plan activities, monitor water quality, and plant woody vegetation to reduce erosion and sediment transfer. "Many of us sheep and beef farmers hadn't really connected with our dairy neighbours because there was no reason to. We now know each other, and it feels like we have a rejuvenated community. The Catchment Group has helped us form relationships I know will last forever." Geordie Eade, Pourakino Catchment Conservation Trust.

Author details

Ian McIvor[1]*, Thomas Mackay-Smith[2] and Raphael Spiekermann[3]

1 Plant and Food Research, Palmerston North, New Zealand

2 Verdantia Research, New Plymouth, New Zealand

3 GeoSphere Austria, Vienna, Austria

*Address all correspondence to: ian.mcivor@plantandfood.co.nz

References

[1] Glade T. Establishing the frequency and magnitude of landslide-triggering rainstorm events in New Zealand. Environmental Geology. 1998;**35**:160-174

[2] Glade T. Landslide occurrence as a response to land use change: A review of evidence from New Zealand. Catena. 2003;**51**:297-314. DOI: 10.1016/S0341-8162(02)00170-4

[3] Goff JR. A chronology of natural and anthropogenic influences on coastal sedimentation, New Zealand. Marine Geology. 1997;**138**:105-117. DOI: 10.1016/S0025-3227(97)00018-2

[4] Green MO. Catchment sediment load limits to achieve estuary sedimentation targets. New Zealand Journal of Marine and Freshwater Research. 2013;**47**:153-180. DOI: 10.1080/00288330.2012.757241

[5] Phillips C, Marden M, Basher LR. Geomorphology and forest management in New Zealand's erodible steeplands: An overview. Geomorphology. 2018;**307**:107-121. DOI: 10.1016/j.geomorph.2017.07.031

[6] Wood PJ, Armitage PD. Biological effects of fine sediment in the lotic environment. Environmental Management. 1997;**21**:203-217. DOI: 10.1007/s002679900019

[7] Blaschke PM, Trustrum NA, Hicks DL. Impacts of mass movement erosion on land productivity: A review. Progress in Physical Geography: Earth and Environment. 2000;**24**:21-52. DOI: 10.1177/030913330002400102

[8] Harmsworth G, Awatere S, Procter J. Meeting water quality and quantity standards to sustain cultural values. In: 21st Century Watershed Technology Conference and Workshop Improving Water Quality and the Environment Conference Proceedings, 3-6 November 2014, University of Waikato, New Zealand. Presented at the 21st Century Watershed Technology Conference and Workshop. New Zealand: American Society of Agricultural and Biological Engineers, The University of Waikato; 2014. pp. 1-8

[9] Dominati E, Mackay A, Lynch B, Heath N, Millner I. An ecosystem services approach to the quantification of shallow mass movement erosion and the value of soil conservation practices. Ecosystem Services. 2014b;**9**:204-215. DOI: 10.1016/j.ecoser.2014.06.006

[10] Dymond JR, Davies-Colley RJ, Hughes AO, Matthaei CD. Predicting improved optical water quality in rivers resulting from soil conservation actions on land. Science of the Total Environment. 2017a;**603-604**:584-592. DOI: 10.1016/j.scitotenv.2017.06.116

[11] Fuller IC, Death RG. The science of connected ecosystems: What is the role of catchment-scale connectivity for healthy river ecology? Land Degradation & Development. 2018;**29**:1413-1426. DOI: 10.1002/ldr.2903

[12] Basher L, Spiekermann R, Dymond J, Herzig A, Hayman E, Ausseil A-G. Modelling the effect of land management interventions and climate change on sediment loads in the Manawatu-Whanganui region. New Zealand Journal of Marine and Freshwater Research. 2020;**54**:490-511. DOI: 10.1080/00288330.2020.1730413

[13] Zabarte-Maeztu I, Matheson FE, Manley-Harris M, Davies-Colley RJ, Hawes I. Fine sediment effects on seagrasses: A global review, quantitative

synthesis and multi-stressor model. Marine Environmental Research. 2021;**171**:105480. DOI: 10.1016/j.marenvres.2021.105480

[14] Jansson MB. A global survey of sediment yield. Geografiska Annaler: Series A, Physical Geography. 1988;**70**:81-98. DOI: 10.1080/04353676.1988.11880241

[15] Basher LR. Erosion processes and their control in New Zealand. In: Dymond JR, editor. Ecosystem Services in New Zealand – Conditions and Trends. Lincoln, New Zealand: Manaaki Whenua Press; 2013

[16] Hicks DM, Hill J, Shankar U. Variation of suspended sediment yields around New Zealand: The relative importance of rainfall and geology. IAHS-AISH Publications. 1996;**236**:149-156

[17] Hicks DM, Shankar U, Mckerchar AI, Basher L, Lynn I, Page M, et al. Suspended sediment yields from New Zealand rivers. Journal of Hydrology New Zealand. 2011;**50**:81-142

[18] Neverman AJ, Donovan M, Smith HG, Ausseil A-G, Zammit C. Climate change impacts on erosion and suspended sediment loads in New Zealand. Geomorphology. 2023;**427**:108607. DOI: 10.1016/j.geomorph.2023.108607

[19] Crozier MJ. Multiple-occurrence regional landslide events in New Zealand: Hazard management issues. Landslides. 2005;**2**:247-256. DOI: 10.1007/s10346-005-0019-7

[20] Dymond JR, Herzig A, Basher L, Betts HD, Marden M, Phillips CJ, et al. Development of a New Zealand SedNet model for assessment of catchment-wide soil-conservation works. Geomorphology. 2016;**257**:85-93. DOI: 10.1016/j.geomorph.2015.12.022

[21] Douglas G, McIvor IR, Manderson AK, Koolaard JP, Todd M, Braaksma S, et al. Reducing shallow landslide occurrence in pastoral hill country using wide-spaced trees. Land Degradation & Development. 2013;**24**:103-114. DOI: 10.1002/ldr.1106

[22] Phillips C, Marden M. Reforestation schemes to manage regional landslide risk. In: Glade T, Anderson MG, Crozier MJ, editors. Landslide Hazard and Risk. Chichester, England: John Wiley & Sons, Ltd.; 2005. pp. 517-547

[23] Phillips C, Hales T, Smith H, Basher L. Shallow landslides and vegetation at the catchment scale: A perspective. Ecological Engineering. 2021;**173**:106436. DOI: 10.1016/j.ecoleng.2021.106436

[24] Dodd MB, McDowell RW, Quinn JM. A review of contaminant losses to water from pastoral hill lands and mitigation options. Hill country – Grassland Research and Practice Series. 2016;**16**:137-148

[25] MacKay AD. Impacts of intensification of pastoral agriculture on soils: Current and emerging challenges and implications for future land uses. New Zealand Veterinary Journal. 2008;**56**:281-288

[26] McDowell RW, Larned ST, Houlbrooke DJ. Nitrogen and phosphorus in New Zealand streams and rivers: Control and impact of eutrophication and the influence of land management. New Zealand Journal of Marine and Freshwater Research. 2009;**43**:985-995

[27] Parfitt RL, Dymond J, MacKay A, Gillingham A, Houlbrooke D, McDowell R, et al. Sources of P in the

Manawatu River and implications for the ONE PLAN. In: Currie LD, Yates LJ, editors. Carbon and Nutrient Management in Agriculture. Palmerston North, New Zealand: Occasional Report No. 21. Fertilizer and Lime Research Centre, Massey University; 2008. pp. 515-524

[28] Donnison A, Ross C, Thorrold B. Impact of land use on the faecal microbial quality of hill country streams. New Zealand Journal of Marine and Freshwater Research. 2004;**38**:845-855

[29] Rosser BJ, Ross CW, C.W. Recovery of pasture production and soil properties on soil slip scars in erodible siltstone hill country, Wairarapa, New Zealand. New Zealand Journal of Agricultural Research. 2011;**54**(1):23-44. DOI: 10.1080/00288233.2010.535489

[30] Peart R, Woodhouse C. Restoring Te Pātaka o Rākaihautu Banks Peninsula. Environmental Defence Society; 2021. 84 p. Available from: https://www.eds.org.nz/ourwork/publications/reports/restoring-banks-peninsula/

[31] Bargh BJ. Output of water, suspended sediment and phosphorus and nitrogen forms from a small forested catchment. New Zealand Journal of Forestry Science. 1977;**7**(2):162-171

[32] Bargh BJ. Output of water, suspended sediment, and phosphorus and nitrogen forms from a small agricultural catchment. New Zealand Journal of Agricultural Research. 1978;**21**(1):29-38. DOI: 10.1080/00288233.1978.10427380

[33] Quinn JM, Stroud MJ. Water quality and sediment and nutrient export from New Zealand hill-land catchments of contrasting land use. New Zealand Journal of Marine and Freshwater Research. 2002;**36**(2):409-429. DOI: 10.1080/00288330.2002.9517097

[34] Pedley DO. Natural Regeneration of Woody Vegetation in Pastoral Hill Country: A Case Study of Oashore Station, Banks Peninsula. Unpublished Masters thesis. New Zealand: Lincoln University; 2022. Available from: https://researcharchive.lincoln.ac.nz/bitstream/handle/10182/15025/Pedley_Masters.pdf

[35] Borda-Niño M, Ceccon E, Meli P, Hernández-Muciño D, Mas J-F, Brancalion PH. Integrating farmers' decisions on the assessment of forest regeneration drivers in a rural landscape of Southeastern Brazil. Perspectives in Ecology and Conservation. 2021;**19**:338-344. DOI: 10.1016/j.pecon.2021.04.001

[36] Bergin DO, Kimberley MO, Marden M. Protective value of regenerating tea tree stands on erosion-prone hill country, east coast, North Island, New Zealand. New Zealand Journal of Forestry Science. 1995;**25**(1):3-19

[37] Smith HG, Spiekermann R, Betts H, Neverman AJ. Comparing methods of landslide data acquisition and susceptibility modelling: Examples from New Zealand. Geomorphology. 2021;**381**:107660. DOI: 10.1016/j.geomorph.2021.107660

[38] Spiekermann RI, Smith HG, McColl S, Burkitt L, Fuller IC. Quantifying effectiveness of trees for landslide erosion control. Geomorphology. 2022a;**396**:1-16. DOI: 10.1016/j.geomorph.2021.107993

[39] Spiekermann RI, Smith HG, McColl S, Burkitt L, Fuller IC. Development of a morphometric connectivity model to mitigate sediment derived from storm-driven shallow landslides. Ecological Engineering. 2022b;**180**:106676

[40] Boffa Miskell (2017), The Mānuka and Kānuka Plantation Guide. Boffa

Miskell, Tauranga. Available from: https://www.trc.govt.nz/assets/Documents/Guidelines/Land-infosheets/Manuka-plantation-guide-landcare-April2017.pdf)

[41] Journeaux P, Praat JP, Handford R, McDonald G. Forestry on Farms: Implications for Farm Sustainability and Regional Impact, AgFirst, Hamilton. Available from: https://www.agfirst.co.nz/projects/forestry-on-farms

[42] Spiekermann RI, McColl S, Fuller I, Dymond J, Burkitt L, Smith HG. Quantifying the influence of individual trees on slope stability at landscape scale. Journal of Environmental Management. 2021;**286**(112194):1-18. DOI: 10.1016/j.jenvman.2021.112194

[43] Mackay-Smith TH, Burkitt L, Reid J, López IF, Phillips C. A framework for reviewing Silvopastoralism: A New Zealand Hill country case study. Land. 2021;**10**(12):1386. DOI: 10.3390/land10121386

[44] Mackay-Smith TH, López IF, Burkitt LL, Reid JI. Kānuka trees facilitate pasture production increases in New Zealand Hill country. Agronomy. 2022;**12**(7):1701. DOI: 10.3390/agronomy12071701

[45] Benavides R, Douglas GB, Osoro K. Silvopastoralism in New Zealand: Review of effects of evergreen and deciduous trees on pasture dynamics. Agroforestry Systems. 2009;**76**(2):327-350. DOI: 10.1007/s10457-008-9186-6

Chapter 9

Prediction of Internal Soil Erosion in Hydraulic Works

Benaissa Kissi, Chafik Guemimi, Miguel Angel Parron Vera, Maria Dolores Rubio Cintas and Rachid Elkhayma

Abstract

Hydraulic structures built on land, such as dams, are numerous and essential as their purpose is to protect people and property (dikes and levees), generate electricity, or create water reserves (dams). Soil erosion, known as hydraulic foxing, is a complex phenomenon which, in its ultimate stage, produces insidious leakage of fluids beneath hydraulic infrastructures known as pipes, and is the main cause of their failure. The HET pipe erosion test is commonly used to quantify the rate of pipe erosion. In this work, the hole erosion test is modeled using ANSYS Fluent software. The aim is to predict the soil erosion rate during the hole erosion test in order to predict the phenomenon of hydraulic foxing within hydraulic structures. The renormalization group theory-based $k - \varepsilon$ turbulence model equations are used. This modeling makes it possible to describe the effect of the clay concentration in flowing water on erosion. Contrary to the usual one-dimensional models, the modeling proposed in this study shows that erosion is not uniform along the entire length of the hole. In particular, clay concentration was found to significantly increase the erosion rate.

Keywords: erosion, piping, hole erosion test, computational fluid dynamics, numerical modeling, embankment dam, risk analysis

1. Introduction

Internal erosion is one of the main causes of instability of earthen hydraulic structures (dike, levee, dam, etc.). The disorders observed on recent structures underline the need for a better understanding and quantification of the phenomena that govern internal erosion. The entrainment and transport of grains by the internal flows affect the granulometric distribution and modify the porosity, as well as the mechanical and hydraulic characteristics.

Dike failures are much more numerous than those of dams due, on the one hand, to the variability of hydraulic solicitations and, on the other hand, to the length and heterogeneity and sometimes the age of dykes and levees, which make monitoring and maintenance difficult.

Numerous dam failures have occurred worldwide, some of them reported by Foster et al. [1]. The main cause of these failures has been identified as being linked to a channeling phenomenon that occurred in the foundation soil or in the dam structure. To ensure the viability of hydraulic infrastructures, it is essential to take into account the vulnerability of the soil to infiltration [2, 3]. Normally, in the case of unconsolidated soils, which are generally made up of slightly cohesive sand particles, we find that water flow velocity plays a very important role in the erosion phenomenon that can occur. Interpreting and understanding the underlying mechanisms and quantifying the effects of relevant variables on this erosion phenomenon is of great practical importance. Soil erosion due to liquid flow can be modeled using a variety of approaches. These include continuum-based models and discrete models, which use certain parameters that are calibrated using laboratory tests or field observations to predict when internal soil erosion begins to occur and the expected erosion rate. Several models for predicting soil erosion rates at the solid/fluid interface have been developed in the literature [3, 4]. One of the most important tests used to predict erosion is the tube erosion test (HET). A model for interpreting the HET with a constant pressure drop has been developed by Bonelli and Brivois [4]. This model provides a characteristic erosion time that depends on the initial hydraulic gradient and the soil erosion rate.

One of the aims of this work is to describe the turbulent two-phase fluid flow that causes erosion inside the porous soil sample, taking into account the influence of the variable clay concentration in the flow fluid. A computational fluid dynamics (CFD) approach will be used to study the shear stress developing at the water/soil interface, which represents the main mechanical action at the origin of surface erosion.

2. Internal erosion processes

It has been found that the soil fractions considered most susceptible to erosion are coarse silts and relatively uniform fine sands.

Cohesive soils, such as clays, are more resistant to erosion as long as chemical bonds are not destroyed [5]. It also seems that certain central materials of glacial origin are particularly sensitive to internal erosion.

According to [6], four conditions must be met for internal erosion and channeling to occur. These conditions are as follows:

There must be a seepage path and a water source.

There must be erodible material within the flow path, and this material must be transported by an infiltration flow.

There must be an unprotected outlet from which the eroded material can escape.

For a channel to form, the transported material, or the material directly above it, must be capable of forming and supporting the channel's "roof."

In an earth and rockfill dam with a central core, there are mainly three processes [6] that can initiate channelization: backward erosion, concentrated leakage, and suffusion. Backward erosion is initiated at the point where the seepage exits, and erosion progresses progressively backward to form a channel. Concentrated seepage begins with a crack or soft zone emanating from the water source to an exit point (downstream). Erosion continues progressively along the walls of the erosion hole, intensifying the concentrated leak. Suffusion is the process by which fine soil particles are carried away or eroded by the voids formed by coarser particles. This phenomenon can be avoided if the soil has a well-spaced granulometric distribution and sufficiently small voids. Soils are said to be internally unstable if suffusion occurs, and internally stable if the particles are not eroded by the infiltration flow.

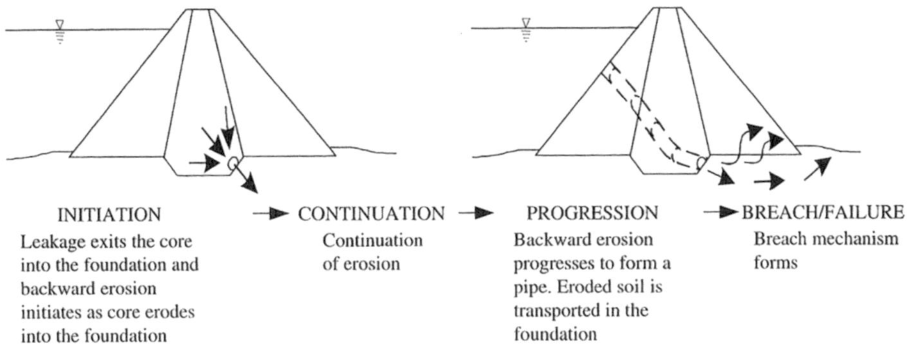

Figure 1.
Conceptual model for development of failure by piping from the embankment into the foundation [7].

Piping can occur in the backfill through the foundation and from the backfill into the foundation. Conceptual piping failure development models for these three cases are shown in **Figure 1**. In addition, a failure path diagram for piping failure through backfill is shown in **Figure 1**. Similar failure diagrams for failure by piping through the foundation and backfill in the foundation can be found in [7] or [8].

3. CFD modeling approach of erosion in the HET

The turbulence of the water flowing inside the HET sample hole is modeled using Fluent software. Fluent is a general-purpose CFD code that has been applied to various problems in the fields of fluid mechanics and heat transfer. The code has been validated in numerous studies. Fluent is, especially, appropriate for the complex physics involved in heat and mass transfer and considers mixtures by modeling each fluid phase independently or as a homogenized medium [9].

Flow taking place inside the hole is turbulent. To perform realistic simulation of turbulence, the exact instantaneous Navier-Stokes governing equations are habitually time-averaged or ensemble-averaged. The obtained averaged equations contain further unknown variables, and turbulence models are introduced in order to determine them in terms of known quantities. Various turbulence models have been proposed in the literature; however, there is no single turbulence model, which could be applied for all classes of problems [10]. The choice of a pertinent model for a given problem will depend on the actual physics of the flow, the degree of accuracy required, and the computational cost tolerated. Fieldview Reference Manual [9] gives a detailed discussion on how to perform at best the appropriate choice of a turbulence model. Among the various models, the standard $k - \varepsilon$ model, which was proposed first by Launder and Spalding [11], has become the most popular one when dealing with practical engineering flow calculations. This model relies on phenomenological considerations and integrates empiricism to perform closure of turbulence equations.

Improvements of the standard $k - \varepsilon$ model, such as the RNG $k - \varepsilon$ model, have been made by [12]. This model was derived by using a rigorous statistical technique called renormalization group theory (RNG) [13].

4. RNG based k-ε erosion model equations for the HET

The RNG $k - \varepsilon$ model differs from the standard model by the special form of the transport equations, which contain the additional term R_ε. These equations write

$$\frac{\partial k}{\partial t}+\frac{1}{r}\frac{\partial(rku)}{\partial r}+\frac{\partial(kv)}{\partial z}=\frac{1}{r}\frac{\partial}{\partial r}\left(\alpha\mu_t r\frac{\partial k}{\partial r}\right)+\frac{\partial}{\partial z}\left(\alpha\mu_t\frac{\partial k}{\partial z}\right)+\alpha\mu_t\frac{k}{r^2}+\mu_t S^2-\varepsilon \quad (1)$$

$$\frac{\partial \varepsilon}{\partial t}+\frac{1}{r}\frac{\partial(r\varepsilon u)}{\partial r}+\frac{\partial(\varepsilon v)}{\partial z}=\frac{1}{r}\frac{\partial}{\partial r}\left(\alpha\mu_t r\frac{\partial \varepsilon}{\partial r}\right)+\frac{\partial}{\partial z}\left(\alpha\mu_t\frac{\partial \varepsilon}{\partial z}\right)+\alpha\mu_t\frac{k}{r^2}+C_{1\varepsilon}\frac{\varepsilon}{\rho k}-C_{2\varepsilon}\frac{\varepsilon^2}{\rho k}-\frac{1}{\rho}R_\varepsilon \quad (2)$$

With

$$R_\varepsilon=\frac{C_\mu\rho S^3k^2\varepsilon(\eta_0\varepsilon-Sk)}{\eta_0\left(\varepsilon^3+\beta S^2k^3\right)} \quad (3)$$

$$S^2=2\left(\frac{\partial u}{\partial r}\right)^2+2\left(\frac{\partial v}{\partial z}\right)^2+\left(\frac{\partial v}{\partial r}+\frac{\partial u}{\partial z}\right)^2 \quad (4)$$

where u and v are average radial and axial flow velocities, r and z are axial and radial coordinates, t is time, ε the rate of dissipation of turbulent kinetic energy, k is the turbulent kinetic energy, ρ is the fluid density, μ_t the total kinematics viscosity, α is the inverse effective Prandtl number for both k and ε, and $C_{1\varepsilon}$, $C_{2\varepsilon}$, C_μ, η_0 and β are constants.

To calculate the tangential shear stress distribution on the cylinder inner wall, boundary conditions must be introduced. When the shear stress is calculated using ANSYS Fluent software, the classical linear erosion law is used to estimate the erosion rate. This law states the estimated erosion rate, defined as the mass departure of particles due to erosion per unit time and per unit area, is expressed by the following formula: $\dot{\varepsilon}_{er}=c_{er}(\tau-\tau_{cr})$ where c_{er} and τ_{cr} are constant depending on the considering soil material. The rate $\dot{\varepsilon}_{er}$ can be related to time variation of local radius by $\dot{\varepsilon}_{er}=\rho_d dR/dt$ where ρ_d represents the dry density of the soil and R the inner radius of the hole. The law of erosion states that the rate of erosion ε_{er} is proportional to the shear stress, exceeding the critical shear τ_{cr} for which erosion begins [14].

5. Interpretation and analysis of results

The fluid is assumed to be axisymmetric, extending over a length of 0.117 m in the axial direction z and 0.03 mm in the radial direction r. The model is designed so that inlet pressure is on the left and outlet pressure is on the right (**Figure 2**).

Figures 3–6 shows shear stress that develops at the soil sample interface with flowing flow for pressure P = 3726 (Pa).

Figure 7 shows that for two clay concentrations, the calculated erosion rate value is 10^{-6} kg/s. This erosion rate value is obtained by integrating the erosion law over the entire length of the sample hole and multiplying the result by the initial hole circumference. The erosion constants used are: $c_{er}=5.6\times10^{-4}s/m$ and $\tau_{cr}=7,1Pa$.
These correspond to a specific soil sample containing 50% kaolinit clay and 50% sand that was tested as reported in Pham [12].

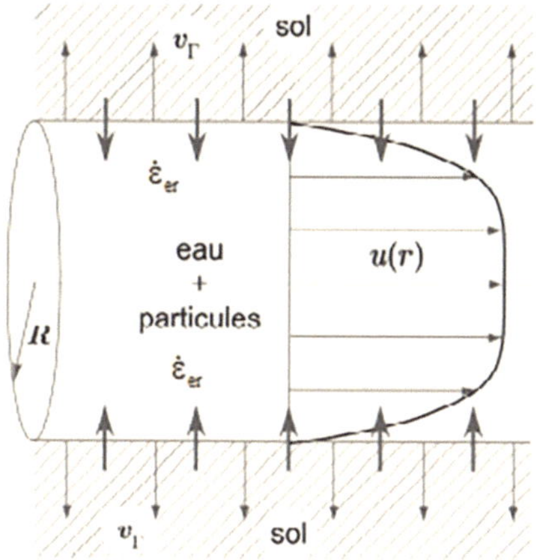

Figure 2.
Geometry of the HET tube.

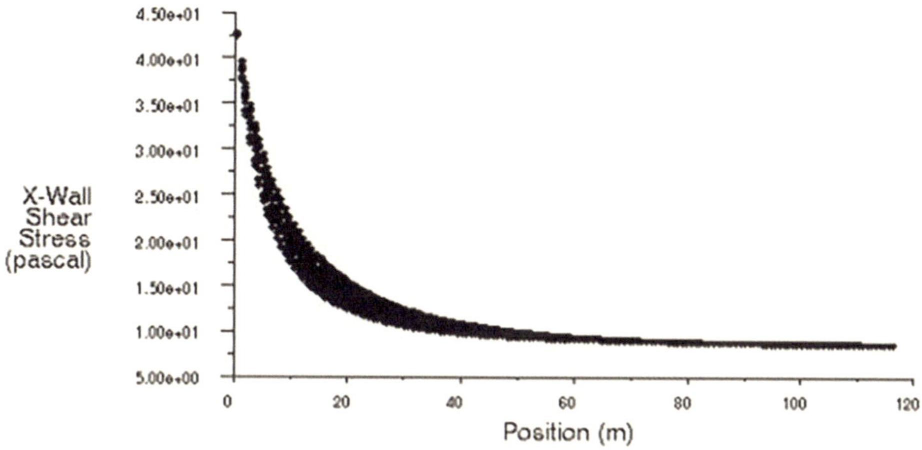

Figure 3.
X-Wall shear stress in pascal.

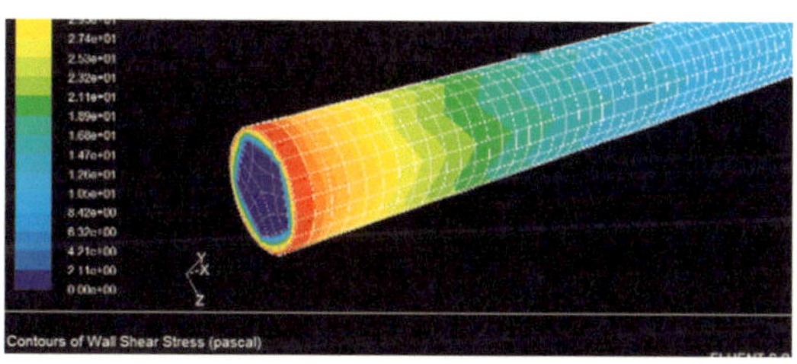

Figure 4.
X-wall shear stress mapping in the solid/fluid interface.

Figure 8 gives the y + value for the applied pressure pinlet = 3726 Pa and the four concentrations as function of the axial coordinate.

Figure 9 gives, for the three applied hydraulic gradients and the four concentrations, curves of the axial velocity at the axis of symmetry as function of the axial coordinate.

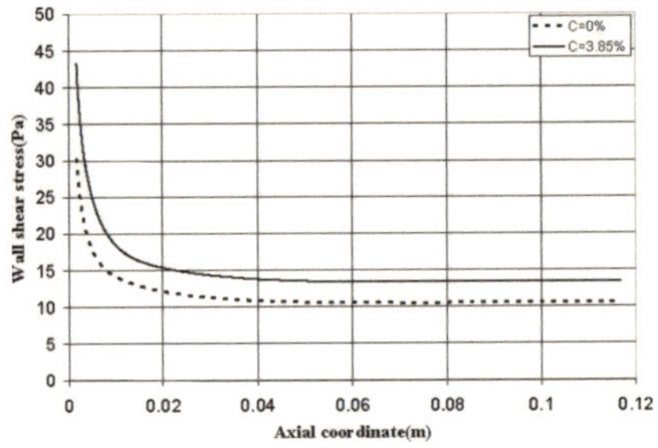

Figure 5.
Wall-shear stress obtained for P = 3726 Pa as function of clay concentration.

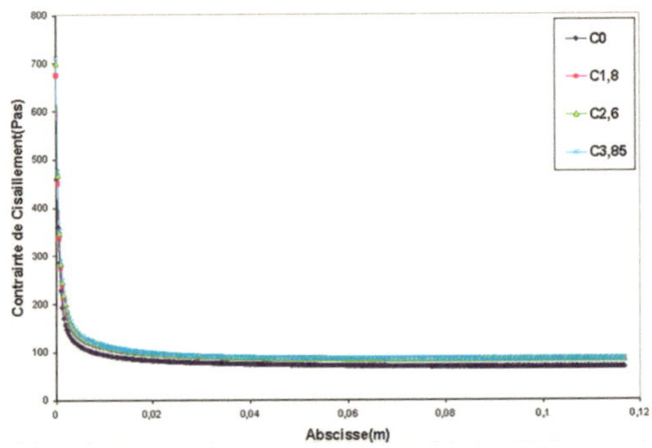

Figure 6.
Shear stress at the interface for the case of $p_{inlet} = 3726$ *Pa as a function of clay concentration.*

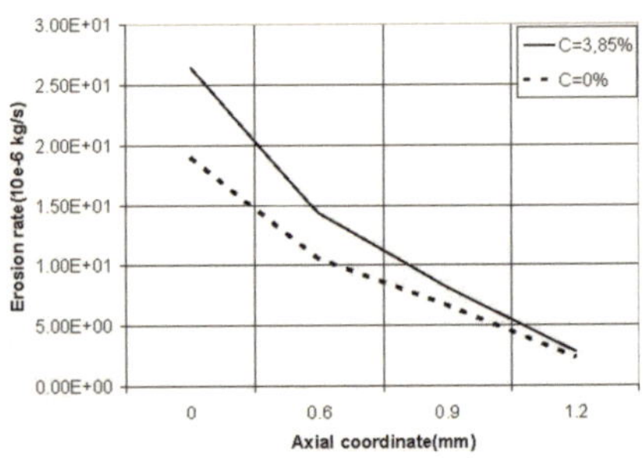

Figure 7.
Erosion rate as function of the clay concentration.

Figure 10 shows the erosion rate as a function of clay concentration for the three pressure gradients. (10^{-6} kg/s).

The results obtained, as shown in **Figure 10**, indicate that the erosion rate increases with clay concentration and with axial coordinate. This contrasts with the one-dimensional EHF model, for which the erosion rate does not depend on the axial coordinate.

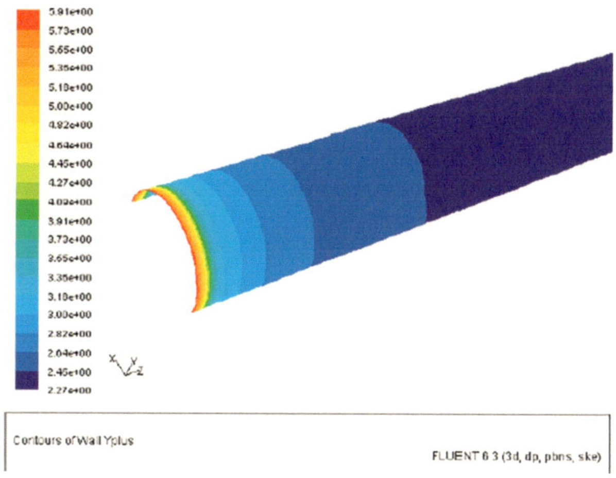

Figure 8.
Contours of wall Yplus on wall.

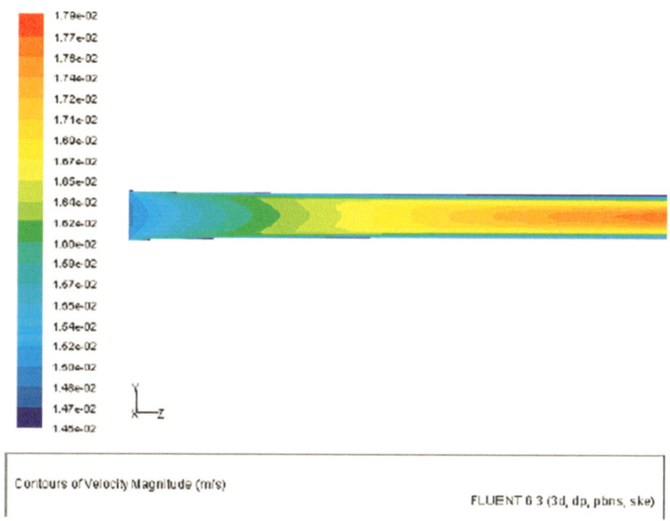

Figure 9.
Contours of velocity magnitude on $x = 0$.

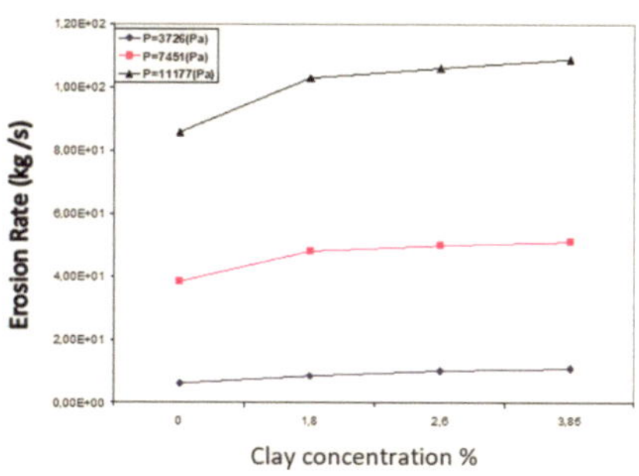

Figure 10.
Erosion rate as a function of clay concentration for the three pressure gradients $(10^{-6} kg/s)$.

6. Artificial intelligence (AI) in civil engineering

The application of artificial intelligence (AI) to civil engineering refers to the use of computer systems to simplify and automate the design and construction processes of civil engineering hydraulic structures.

In its broadest sense, artificial intelligence (AI) represents the scientific discipline concerned with the study, design, and implementation of technologies capable of imitating the cognitive abilities of a human being.

Voice assistants, home automation systems, search engines, satellite navigators, etc. are common examples of intelligent systems, capable of understanding the interests and demands of users and providing responses tailored to their needs.

AI has long been used successfully in many fields: from manufacturing to medical diagnostics, from e-commerce to video games, and more. Today, more than ever, we are witnessing the development and deployment of intelligent systems in the field of civil engineering.

Thanks to these systems, which are able to learn from their mistakes and carry out activities similar to those of humans, it is now possible to solve many of the problems associated with the construction of buildings and infrastructures.

Before analyzing the various applications of AI in civil engineering, let us find out what learning techniques this technology is based on.

Unlike conventional simulation software, machines equipped with artificial intelligence have the ability to:

- perceive the world around them and collect useful data and information.
- understand perceived reality, by logically linking collected information.
- perform autonomous operations (computerized or mechanical), deciding to act without any command from a human being.
- learn from the results of its actions, continuously improving, and learning from its mistakes.

Machine learning is one of the fundamental characteristics of intelligent systems used in civil engineering. This capability can be developed using the following techniques:

- Evolutionary computation: This is a learning technique based on the concepts of evolutionary biology. This technique is implemented on computer systems to solve complex problems characterized by too many variables to be handled by traditional algorithms. Evolutionary computation is widely used in civil engineering to solve optimization problems and automate the production of civil engineering projects.
- Artificial neural networks: Mathematical models composed of artificial neurons that reproduce the behavior of the human brain. Neural networks enable computer programs to recognize patterns and solve common problems. In engineering, they are used to improve decision-making processes, make predictions, perform data analysis, monitor structures, control the movement of robots, etc.

- Fuzzy systems: The term "fuzzy" refers to the ability of these systems to process imprecise or vague data, using a mode of reasoning close to that of humans (who do not always adopt rigorous and well-defined logic). Fuzzy logic helps computer systems used in civil engineering to deal with incorrect inputs and outputs and to model construction deadlines, costs, and risks. Fuzzy systems are also used to assess the quality of infrastructure projects.

- Expert systems: These are technological applications capable of solving problems in a very specific field, and which can reach, or even exceed, the human performance of an expert operating in that specific sector. Expert systems are mainly used in the construction and geotechnical sectors to analyze the energy consumption of buildings or to carry out geological surveys.

7. Conclusion

In this work, a two-dimensional modeling of the tube erosion test was carried out to predict the erosion phenomenon within hydraulic structures. Unlike, the first models produced by several authors, which are essentially one-dimensional and the two-dimensional modeling of this phenomenon showed that the tangential shear stress along the interface wall between the water and the body of the structure is not uniform. By applying a linear erosion law, we were able to predict nonuniform erosion along the entire length of the interface.

The study of the effect of clay concentration showed that it has a very significant effect on the evolution of wall shear stress, and therefore, in turn, affects the surface erosion that develops at the soil sample fluid interface, particularly at the end of the hole outlet where it is at its maximum. The results obtained have enabled us to understand qualitatively why the eroded profile of the hole wall, as observed during the HET experiment, is not uniform.

Author details

Benaissa Kissi[1]*, Chafik Guemimi[1], Miguel Angel Parron Vera[2], Maria Dolores Rubio Cintas[2] and Rachid Elkhayma[1]

1 National Graduate School of Arts and Crafts, Hassan II University, Morocco

2 Industrial and Civil Engineering Department, Polytechnic High School of Algeciras, University of Cadiz, Spain

*Address all correspondence to: benaissa.kissi@gmail.com

References

[1] Foster MA, Fell R, Spannangle M. The statistics of embankment dam failures and accidents. Canadian Geotechnical Journal. 2000;**37**(5):100-1024

[2] Fjar E, Holt RM, Horsrud P, Raaen AM, Risnes R. Petroleum Related Rock Mechanics. Revised ed. Amsterdam: Elsevier; 2004

[3] Lachouette D, Golay F, Bonelli S. One dimensional modelling of piping flow erosion. C. R. Mecanique. 2008;**336**: 731-736

[4] Bonelli S, Brivois O. The scaling law in the hole erosion test with a constant pressure drop. International Journal of Numerical Methods in Engineering. 2008;**32**:1573-1595

[5] Srbulov M. Estimation of soil internal erosion potential. Computers and Geotechnics. 1988;**6**:265-276

[6] Fell R, MacGregor P, Stapledon D, Bell G. Geotechnical Engineering of Dams. Leiden: Balkema; 2005. ISBN: 041536440x

[7] Foster MA. The probability of failure of embankment dams by internal erosion and piping [PhD thesis]. Sydney: School of Civil and Environmental Engineering, The University of New South Wales; 1999

[8] Skoglund M, Solvik Ø. External and internal erosion in rockfill dams. International Journal on Hydropower & Dams. 1995;**2**(3):44-47

[9] Fieldview reference manual, software release version 10, intelligent light. 2004

[10] Escue A, Cui J. Comparison of turbulence models in simulating swirling pipe flows. Applied Mathematical Modelling. 2010. DOI: 10.1016/j.apm.2009.12.018

[11] Launder BE, Spalding DB. Lectures in Mathematical Models of Turbulence. London, England: Academic Press; 1972

[12] Pham TL. Erosion et dispersion des sols argileux par un fluide. In French [PhD thesis] Order number: D.U. ED: 430. Ecole Nationale des Ponts et Chaussées, Paris, France. 2008

[13] Yakhot V, Orszag SA. Renormalization group analysis of turbulence: I-basic theory. Journal of Scientific Computing. 1986;**1**(1):1-51

[14] Kissi B, Parron Vera MA, Maria Dlolores R, Dubujet P, Khamlichi A, Bezzazi M, et al. Predicting initial erosion during the hole erosion test by using turbulent flow CFD simulation. Applied Mathematical Modelling. 2012; **36**(8):3359-3370

Chapter 10

Using Rainfall Simulators to Design and Assess the Post-Mining Erosional Stability

Ashraf M. Khalifa and Hwat Bing So

Abstract

The mining industry is crucial for global economic growth but faces environmental challenges, especially in designing stable rehabilitated landforms. To tackle these issues, rainfall simulators have been recognized for their value in providing data for erosion modeling and analysis, aiding the development of effective land cover systems for long-term stability. This chapter provides an overview of the theory, specifications, and design principles of rainfall simulators. It explores the detailed design and construction of a well-known model, along with its calibration process ensuring accurate rainfall production and distribution. The chapter also discusses raindrop size distribution and associated kinetic energy calculations. Calibration results demonstrate satisfactory outcomes with Christiansen's uniformity coefficient exceeding 85% and a median raindrop size of 2.15 mm. The device successfully generates desired kinetic energy for simulated rainstorms, crucial for studying soil erosion. Examples highlight the application of rainfall simulators in evaluating erosion stability in Queensland mines. Efforts to construct a soil erosion database for 34 open-cut mines in Queensland using a similar portable rainfall simulator are highlighted. This database contributes to developing user-friendly MINErosion models, providing estimates of soil erosion/deposition at different scales to support the Australian mining sector.

Keywords: soil erosion, rainfall simulation, simulator, erosion modeling, mining, rehabilitation, erosional stability, landform design, land cover, MINErosion, WEPP

1. Introduction

The mining industry plays a crucial role in global economic development, but it often faces environmental challenges, particularly related to rehabilitated landform design and its erosional stability. Constant efforts must be made to develop and build an appropriate landform and cover system that effectively serves its intended purposes, such as supporting vegetation growth and preventing the infiltration of water and oxygen into reactive mine waste. The success of the soil cover system depends on the external surface of the landform remaining intact. If the surface erodes, the cover system's functions are compromised, and it is likely to fail. To evaluate the long-term erosion stability of a constructed landform, erosion rate thresholds must be established to ensure the landform is acceptably resistant to erosion, or "stable."

While the defensible erosion rate thresholds should come close to the term "tolerable or acceptable erosion" rate which was first proposed by Browning et al. [1], and a more thorough review of tolerable values was carried out by the Soil Science Society of America in 1979; there is still no broad agreement on what we could consider as a tolerable or "acceptable" rate of erosion on a rehabilitated landform.

In general, US soil conservationists have consistently based tolerable soil loss values largely on the natural soil forming rate and on consideration of the maintenance of soil productivity, although accepting that other factors may be important in some situations. The US soil conservation agencies have typically used tolerable erosion values of < 11.2 t/ha/y for deep fertile soils, and < 4.5 t/ha/y for shallow agricultural soils. However, the fact that these values have been quantified under the American agricultural soil conditions must be kept in mind; that makes it doubtful to blindly accept these values to judge or establish defensible erosion rate thresholds under the Australian mining conditions. Therefore, the Queensland Department of Mines and Energy previously used a range of 12–40 t/ha/year as a target erosion rate for rehabilitated mine sites [2, 3], which was also impacted by the fact that it should be manageable to levels that do not compromise post-mining land use.

It is also worth mentioning that there are a few Australian studies that have attempted to quantify erosion rates. Lu et al. [4] utilized spatial modeling methods (remotely sensed data) to predict the sheetwash (interrill) and rill erosion all over the Australian continent. They estimated that the average erosion rate is 4.1 ton/ha/year over the Australian continent; however, they stated that "Soil erosion is naturally highly variable. This needs to be recognized when comparing current rates of erosion from one place to another and when the erosion control policies are set. It should be expressed in relation to spatially variability, rather than referring to absolute rates alone or using a single benchmark applied across diverse landscapes."

Since it seems difficult to agree on a specific value for erosion rate threshold, the determination of this threshold value must be done for each mining site independently, to make it conceivable to achieve the required erosion stability; and that which decision makers and governmental regulators work on this matter will accept. The concept that the soil erosion rate threshold for specific mining site should be equal or close to the erosion rate values of the surrounding areas in similar climatic and environmental conditions deserves support and should be applied; the value of this threshold should be also manageable to levels that do not compromise post-mining land use agreed upon with the local community and by the PRC plan.

Rainfall simulators are distinguished research tools that enable us to measure erosion rates for any mine site and its surrounding area in the laboratory or in the field with high accuracy and efficiency; the erodibility values of soil and materials that will be used as land cover can also be measured accurately. Therefore, the rainfall simulators can provide us with some important data necessary for analysis and erosion modeling processes to help us design the best land cover system that could achieve long-term erosional stability.

2. Rainfall simulator: assessing soil erodibility for rehabilitation sites

Rainfall simulators are commonly used as research tools to study soil erosion [5–12], as well as, soil infiltration [13, 14], water runoff and sediment transport [15–17], geotechnical studies especially slope stability and landslides [18–23], and many other related research areas. Rainfall simulators can produce unique data that are vital for calibration and validation purposes of empirical, conceptual, or process-based rainfall-runoff-sediment transport mathematical models [24].

The credit for designing and building the first rainfall simulator goes to Nichols and Sexton [25] who used their first rainfall simulator (spray-bar rainfall simulator) to study soil erosion and measure the infiltration rate. By the mid-1950s of the last century, the use of rain simulators expanded steadily in experiments related to soil erosion, and the rapid technological development contributed to the introduction of many improvements to its first designs to avoid many issues that affect the performance, results, credibility, and the feasibility of these machines [26].

The main purpose of a rainfall simulator (RFS) is to generate and create an artificial rainstorm with precise specifications in terms of the duration and intensity of the rainfall, as well as in some way the size distribution of the droplets, and its kinetic energy. The ability to control the physical characteristics of the generated rainstorm makes it possible to keep the climatic factor (rain erosivity) constant while studying other factors that may affect the process of erosion such as soil erodibility, slope factor, or vegetation cover. It can be argued that any success of the RFS design depends entirely on how closely it is simulating natural precipitation conditions with respect to spatial uniformity, raindrop size, raindrop terminal velocity, and kinetic energy.

Although rainfall simulators are frequently utilized in soil erosion experiments, their ability to replicate natural rainfall conditions with precision has been a topic of concern. Many studies have been carried out to evaluate the reliability of results obtained from rainfall simulators and their usefulness in modeling soil erosion processes [27–31]. Among the important initiatives in this domain was the PLPEWC "Post-mining Landscape Parameters for Erosion and Water Quality Control" project, which was financially supported by ACARP (the Australian Coal Association Research Project) between 1992 and 1998.

The project performed a range of experiments, including laboratory rainfall simulation, field plots, and catchments (**Figure 1**).

The experimental approaches adopted were designed to measure the basic erosion parameters at the different scales. A large amount of data has been collected on 34 spoil and soil materials from 16 mines in Central Queensland, as well as 9 years of field plot and field catchment data [7, 32]. The data collected from those different experimental approaches/scales studies proved that although the need for field plots and catchment flumes studies still exists, the results obtained from laboratory rainfall simulators showed reliability so that their results can be used in modeling soil erosion

Figure 1.
Range of experimental approaches adopted to determine soil/spoil erodibility in a previous ACARP (the Australian coal association research project).

with a high degree of accuracy. Moreover, laboratory experiments using rainfall simulators are more manageable than field experiments, because the data on runoff and soil loss can be obtained without waiting for natural rain to happen.

Therefore, rill and interrill erodibilities and slope adjustment factors were measured for these 34 soil/overburden materials on a portable rainfall simulator tilting flume (3 m long × 0.8 m wide, slope adjustable from 0 to 50%) at the University of Queensland Erosion Processes Laboratory. Each material was subjected to 100 mm. h^{-1} rainstorm for 30 min (equivalent to a 1-in-20-year event in Central Queensland) at 20% slope, followed by slopes of 5, 10, 15, and 30% for 15 min each. At these simulated rainfall intensities, a steady state was quickly formed. The data from these measurements and the derived parameters were used to develop the MINErosion V3.x model [33], which successfully estimates field scale erosion rates on simple linear hillslopes with various combinations of slope gradients and lengths. MINErosion 3 can also be used effectively to simulate multiple field plot experiments on a computer, based on a few measurements made on a tilting flume-rainfall simulator facility in the laboratory. MINErosion 3.4 cannot be used to predict sediment yield from a watershed with complex topography in terms of slope steepness and flow pathways. However, it is necessary and desirable to be able to estimate off-site sediment discharges from these rehabilitated post-mining landscapes. For this purpose, MINErosion 4 was developed, which combines the MINErosion 3.4 model and a geographic information system (GIS) package (ESRI ArcGIS 10.3 or the freeware QGIS 3.16), to estimate erosion rates and sediment movement and delivery from these constructed post-mining landscapes. Both MINErosion 3 and 4 demonstrated the opportunities and the value of using the rainfall simulators at the mining sites to model and assess the erosional stability, which should be proven achievable under the given circumstances as it is one of the main considerations of the landform design report within the progressive rehabilitation and closure (PRC) Plan.

Subsequently, the effectiveness of using rainfall simulation as a method for obtaining erodibility information for other soil erosion models such as Areal Nonpoint Source Watershed Environmental Response Simulation (ANSWERS), Chemicals, Runoff, and Erosion from Agricultural Management Systems (CREAMS), and Water Erosion Prediction Project (WEPP) was recognized. Loch, Silburn [34], Silburn and Connolly [35], and Silburn and Loch [36] achieved accurate predictions of erosion under field conditions by utilizing parameters derived from rainfall simulation using these models. Nevertheless, Silburn and Loch [36] emphasized the significance of ensuring that the erosion processes happening on rainfall simulator plots were identical to those occurring in field areas in order to obtain reliable predictions.

3. Rainfall simulator: specifications and design

Although there are hundreds of different designs of rain simulators [6, 7, 15, 17, 22, 24, 37–53], each striving to get closer and closer to simulating the distribution of natural rain with varying degrees of success, rain simulators can be divided into two main groups [54] based on the way raindrops are generated:

- Non-pressurized rainfall simulators.
- Pressurized rainfall simulators (nozzle type).

The earliest versions of the rainfall simulators belong to the non-pressurized category, where drop-forming mechanisms completely rely on passing the water through a perforated pipe, hanging yarns, or an array of syringe needles which form the droplets, then the droplets are left free to fall under the impact of gravity from a height of not less than 9.1 m to ensure the droplets reach the required terminal velocity, which should be almost equal to the one of the natural rainfall droplets. This type of rainfall simulator has always suffered, due to the long distance between the raindrop generator and the flume surface, from the effect of wind on the falling droplets which make it a necessity to install a huge wind shield (**Figure 2**). The huge structure (the frame and wind shield) makes it impossible to use this type of rainfall simulator in field experiments (lack of portability). Later, the second type of rainfall simulator appeared, which depended on nozzles and pressurized water flow system, and it achieved a widespread popularity—at the expense of the first type—due to it being more portable and usable in field experiments as it is smaller in size (no more than 2–3 m high, **Figure 3**) and less expensive to build and run. Nevertheless, due to the pressurized nature of the simulator which produced a high-intensity spray, rainfall intensity is usually controlled by applying the water intermittently.

In this section, we will describe the features, structure, and calibration processes of a modified version of Queensland Department of Primary Industries (QDPI) rainfall simulator (as depicted in **Figure 3**), The device has undergone testing by numerous researchers [7, 55]. Queensland Department of Primary Industries (QDPI) rainfall simulator has been constructed multiple times in various locations throughout Queensland and beyond, with minor variations between each iteration. As a result, it can be replicated with ease by others. The authors of this chapter constructed the Griffith University version of the model, known as the Port-RFS, which will be presented and discussed here. This rainfall simulator is characterized by the following specifications:

- High portability, which means it should be lightweight and easy to assemble, as it should be easy to transport from one site to another.

- Efficiency in water usage (as the available supplies of acceptable quality water were sometimes very limited in some locations).

Figure 2.
The Griffith University tilting flume simulated rainfall (GUTSR). The left picture shows the rainfall generator above the flume. The right picture shows a side view of the sloping flume and the wind shield.

Figure 3.
The Griffith University portable rainfall simulator (port-RFS) installation and operation in the field.

- Reliable and ease of fixing and repairing on site; and
- Suitability for operation by a fieldwork team of three or four people.

As a pressurized rainfall simulator (nozzle type), the Port-RFS consists of a structural frame, the rain drop generator system, the water supply unit (tank), and the flume/soil container. The structural frame was constructed from aluminum tubing of 38 mm outside diameter (O.D.) and a 3 mm wall thickness. The bottom sections of some of the upright tubes are bent to form detachable legs to make it possible for the frame to be mounted on a hydraulic-tipper trailer when used inside a closed place/laboratory, so that it can use the slope mechanism of the trailer; as well as make it possible to be pegged to the ground, if the Port-RFS is used in the field (**Figure 3**). In order for the RFS to be transportable, the frame consists of a number of separate parts (12 pieces of metal tubes) that are connected together by nylon joiners that fit inside the end of aluminum tubes. The joiners are machined from solid nylon material and are about 180 mm in length. Locking pins ensure that the vital parts are securely interconnected.

The nozzle boom is made up of a 4 m long aluminum tubing with an outer diameter of 38 mm and a wall thickness of 1.6 mm. It can be shortened to 3 m to allow for mounting on a hydraulic-tipper trailer of the same length. The boom rotates in two plastic bushes (graphite-impregnated) that also prevent lateral movement of the boom. Four male threaded (1/2" BSPT (British Standard Pipe Taper)) unions are welded onto the boom 1 m apart. Check valves are threaded onto the bushes to prevent nozzle flow or dripping when the unit is not in use, and nozzles are fitted to the check valves. The water supply inlet is via a 1¼" BSPT tee fitting attached to the boom, opposite to the row of nozzles. A tapping from this fitting provides a connection for a pressure gauge, so that the operating pressure of the unit can be monitored. The rain drop generator system of Port-RFS consists of three to four VeeJet 80,100 nozzles (depending on the length of the boom) that are installed on the boom, 1 m apart, to cover (overlapped) the flume area (3 m × 1 m) underneath. A McLennan Unipolar Permanent Magnet Stepper Motor 1.8°, 3.8 Nm, 120 V dc, 4.3 A, eight wires with a programmable digital controller to drive and control the oscillating action of the nozzles' boom. The water supply unit consists of Matrix 10-5 VFD: The Ward 10-5304 stainless steel horizontal multistage pump coupled to a 2.2 kW single-phase motor drawing 13 Amp full load current, with 40 mm female BSP inlet and 32 mm

Figure 4.
Excess water collecting trays.

female BSP outlet. Controlled via the SteadyPress variable speed drive unit capable of a maximum flow rate of 200 L/min; the pump was used to feed the system with water from a 200-liter water tank. A solenoid valve and pressure gauge are attached to the water tank unit to help control the water flow rate.

The decision to use VeeJet 80,100 nozzles was made based on previous personal experience of the authors as well as the previous research of Bubenzer [56], Boulange, Malhat [15], and Loch, Robotham [7]. In pressurized rainfall simulators (nozzle type), it is important that the water flow rate provide enough pressure to allow the rain drops to have the capability to reach the required terminal velocity [46], which ensures that the kinetic energies of the generated storm satisfactorily resemble those of an intense natural rainstorm as well as the drop size distribution of erosive storm patterns. The operational sequence of this RFS relies on a continuous flow of water through the nozzles. Excess water that falls outside the soil is recycled via catch trays manufactured from galvanized aluminum plates that are arranged to collect excess water from either side of the individual oscillating nozzle (**Figure 4**).

During operation, the nozzles oscillate through “108°. Of this trek, the middle ≈ 68° applies raindrops onto the flume area below, and 20° is used at either end of the travel for overlapping, then the excess water gathered at the catch trays will go back to the water tank. Changing the frequency of nozzle oscillation using the stepper motor controller board, we can control and adjust the rainfall intensity coming by changing the waiting time and the consecutive sweeps (**Figure 4**). The stepper motor's control system is composed of a DVP-14SS211T2 Programmable Logic Controller circuit board, EM806 Stepper Driver, a 4-button digital switch, and transistor switching circuits. The microcontroller runs in single chip mode using only internal random-access memory (RAM) and electrically erasable, programmable read-only memory (EEPROM) for data and program storage. The 4-button digital switch set the waiting time between consecutive sweeps of the spray manifold in 0.1-s increments. The transistor switching circuit is required to provide the correct voltage and current levels to the stepper motor driver module.

4. Rainfall simulators: Performance and calibration

The characteristics of simulated rainfall are reliant on the type of nozzle employed, the water pressure exerted, and the arrangement and movement pattern of the nozzles through the rotating bar (nozzle boom) [24]. Therefore, the performance of any rainfall simulator must be evaluated for the ability to produce different rainfall intensities (rainfall intensity calibration), the spatial uniformity of the simulated rainfall over the flume area, the rain Drop Size Distribution (DSD), the terminal velocity, and kinetic energy of simulated rainfall events.

4.1 Rainfall intensity calibration

To be effective in soil erosion experiments, a reliable rainfall simulator should have the capability to generate simulated rainfall events with a broad range of rainfall intensities. Typically, the rainfall intensity utilized in soil erosion experiments falls between 50 and 120 mm/hour, although it may be increased to 150 mm/hour for specific experiments or reduced to 30 mm/hour for less severe erosion studies [33]. The majority of present-day portable rainfall simulators, such as the QDPI rainfall simulator, manage the intensity of produced rainfall by manipulating two factors. First, the water flow/pressure from the primary water pump is regulated using a solenoid valve and pressure gauge. Second, the movement pattern of the nozzles is altered by employing a stepper motor controller and driver to modify the sweep and waiting time pattern. To verify the success of the apparatus (the rainfall simulator) in producing rainfall events with the desired intensities for the experiment or intended application, it must be calibrated using one of the following methods:

- Calibration using a flowmeter: This method involves measuring the flow rate of water through the nozzle that produces the raindrops. The flow rate can be measured using a flowmeter, and the rainfall intensity can be calculated by dividing the flow rate by the area of the nozzle.

- Calibration using a tipping bucket rain gauge: This method involves collecting the water that falls from the rainfall simulator using a tipping bucket rain gauge. The rainfall intensity can be calculated by dividing the volume of water collected by the time it took to collect it.

- Calibration using a high-speed camera: This method involves recording the artificial raindrops using a high-speed camera and analyzing the footage to determine the size and velocity of the raindrops. The rainfall intensity can be calculated using the size and velocity of the raindrops and the nozzle area.

- Calibration using a rain gauge array: This method involves setting up a rain gauge array around the rainfall simulator to measure the rainfall intensity at different distances from the simulator. The rainfall intensity can be calculated by analyzing the data collected by the rain gauge array.

It is important to note that the calibration method used will depend on the specific rainfall simulator being used and the accuracy required for the experiment or application. The rainfall intensities generated by the "Port-RFS" were evaluated and adjusted by taking into account the average volume of water collected in the flume area using the pan method. Additionally, the water discharge from each Port-RFS nozzle was measured by enclosing polyvinyl chloride (PVC) tubes around each

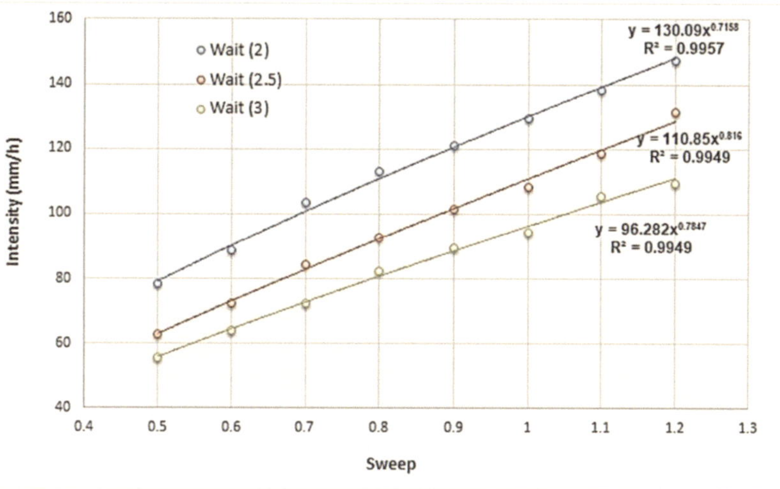

Figure 5.
The Griffith University portable rainfall simulator (port-RFS) calibration curve for the rainfall intensities, relation between rainfall intensity, sweep and waiting times.

nozzle individually and collecting the outflow for a duration of 5 min. To convert the collected data into flow rate in millimeters per hour, the recorded value was multiplied by 12.

Figure 5 shows the calibration of the Port-RFS where rainfall intensity can be controlled at rates from 60 to 150 mm/h using combinations of waiting and sweep periods. It is evident that the Port-RFS has the capability to produce simulated rainfall storms ranging from a minimum intensity of 60 mm/h to a maximum intensity of 150 mm/h. The control of rainfall intensity in the simulated rainstorm was found to be straightforward and efficient using the digital control panel.

4.2 Spatial uniformity over the flume area

Obtaining a uniform distribution of rainfall across each section of the flume is crucial. Failure to achieve this can result in areas that receive more rainfall being more prone to erosion, compromising the accuracy of calculations based on the entire flume area. To ensure uniform rainfall distribution, the grid method is usually used to assess the spatial uniformity of simulated rainfall. It involves superimposing a grid with equidistant points onto the flume area and measuring the amount of rainfall that falls on each point using a rain gauge/graduated beaker. The gathered data for each point are then utilized to compute Christiansen's uniformity coefficient (CUC), as shown in Eq. (1) [57].

$$CUC = 100\left[1 - \left(\frac{\sum(|D_i - D_m|)}{n * D_m}\right)\right] \quad (1)$$

where CUC is the coefficient of uniformity (%); Di is the depth of water in the graduated beakers (cm); Dm is the mean depth of water in rain gages/graduated beakers (cm); and n is the number of rain gages/graduated beakers. When the rainfall

Rainfall intensity (mm/h)	130	120	110	100	90	80	70	60
Uniformity coefficient (%)	89.3	91.8	89.6	90.1	89.6	89.8	89.4	86.55

Table 1.
The calculated uniformity coefficient % for portable rainfall simulator (port-RFS) under different rainfall intensities.

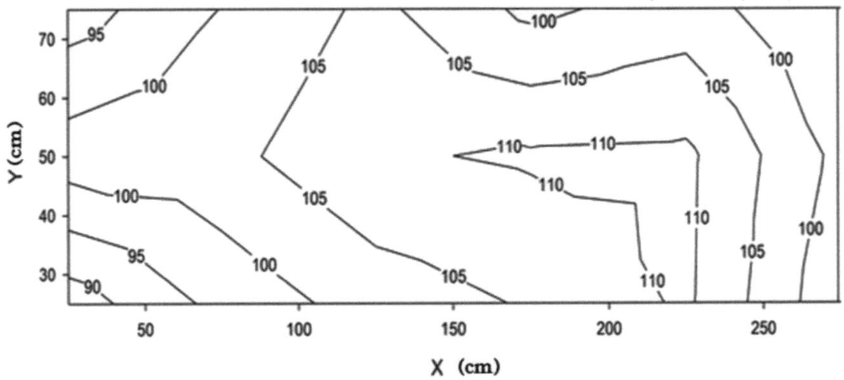

Figure 6.
The spatial distribution of the simulated rainfall over the flume/plot area for the Griffith University portable rainfall simulator (port-RFS) (100 mm/h rainfall intensity).

pattern is more uniform, the CUC value approaches 100%. According to Sousa, Mendes [58], several researchers consider that any CU values above 80.0% are acceptable for the uniformity of the rainfall distribution. However, some other studies have accepted a CUC value of 70% for large plot areas, as demonstrated by Luk, Abrahams [59]. The uniformity coefficients of the Port-RFS at various rainstorm intensities are presented in **Table 1**. All the coefficients exhibit high values, ranging from 86.55% to 91.8%. These values indicate a high level of uniformity across the measured experimental area. **Figure 6** illustrates the distribution of rain intensities generated by the Port-RFS system for an average rain intensity of 100 mm/h across a flume area measuring 3 m × 1 m. The figure demonstrates that the incident intensities vary between 90 and 110 mm/h, with an average uniformity coefficient of 89.8%.

4.3 The drop size distribution (DSD) and kinetic energy (KE)

The ability of any rainfall simulator to generate raindrops that approximate the volumetric size distribution of the droplets that occur during rainstorms in nature is highly influential in our judgment of the efficiency and quality of the rainfall simulator design, as the distribution of grain size over the different classes of drop sizes (volume in mm3) affects the total kinetic energy generated from the simulated rainstorm, whereas the kinetic energy of a single drop is a function of a grain's mass, which is related to its size (volume) as well as its terminal velocity when it hits the ground [49, 60]. In general, the sizes of raindrops in nature range from 0.5 mm in diameter to the large drops associated with heavy rainfall and reaching up to 6 mm in diameter, with median droplet diameter varying depending upon the storm intensity but usually ranging from 2 mm to 3 mm [9, 46].

There are numerous methods and instruments for measuring raindrops, which can be divided into two main groups: manual and automated techniques. These approaches are used to determine the raindrop size distributions and the average size of raindrops for simulated rainfall events.

Manual rain drop measurement techniques include the stain method that involves using dyed absorbent paper to measure the stains left by raindrops [61], the flour pellet method that uses finely sieved flour to create dough pellets from raindrops [62, 63], and the oil immersion method that measures raindrops in a vessel containing oil [64]. While these methods are simple and inexpensive, they are time consuming, the accuracy of the results obtained from it depends on the skills and experience of the researcher, and do not provide immediate data readings.

On the other hand, there are various automated techniques available for measuring raindrops, including but not limited to: the displacement disdrometers [65], the photographic method [66, 67], acoustic disdrometers [68], the radar technique [69, 70], and the optical spectra pluviometers [71].

While the disdrometer method has been particularly successful over the past decade due to its ability to generate a large number of measurements [72, 73] and its efficiency in measuring the impact of raindrop splash on soil detachment [74] and erosion caused by changes in soil cover [75], the old flour method [62] is a widely accepted, standardized test method [46, 49, 63]. Using the flour method, the mean diameter of the raindrops that came out of each examined rain event was measured and could be calculated using Eq. (2).

$$D_r = \sqrt[3]{\left(\frac{6}{\pi}\right) W m_R} \tag{2}$$

Where D_r is the mean raindrops' diameter (mm) and W is the mean weight of the raindrops (mg) and m_R is the ratio of the mass of the raindrop to the mass of the pellet, which is obtained using the flour-calibration line [76].

Using Eq. (3), the kinetic energy of individual raindrops could be determined after the median size distribution (D50) of the rainfall simulator was easily estimated from the previous step:

$$\mathrm{KE} = \tfrac{1}{2}\,\mathrm{mv}^2 \tag{3}$$

where KE is the kinetic energy (Joule); m is the mass (kg) of the raindrop (calculated from the relation between the volume, density, and mass); and v is the terminal velocity (m/s) at which the drop hits the soil surface where the values for examined rainfall intensities by examined rainfall simulator could be obtained from the American Society for Testing and Materials (ASTM) [77] chart that correlates the fall velocity, fall height, and raindrop diameter.

The threshold kinetic energies needed to initiate soil detachment (erosion) by raindrop impact were listed and discussed by Salles, Poesen [78]; they stated that the threshold kinetic energy required to initiate the detachment of soil particles by raindrop impact declines with increasing median grain-sized diameters, starting from 0.001 mm until D50 reaches values near 0.1–0.2 mm. Once D50 becomes larger, the variation in the threshold kinetic energy changes and increases with median grain diameter of the soil.

By utilizing the flour method, the drop size distribution of the Port-RFS was examined, resulting in a median size distribution of 2.15 mm; **Figure 7** presents the

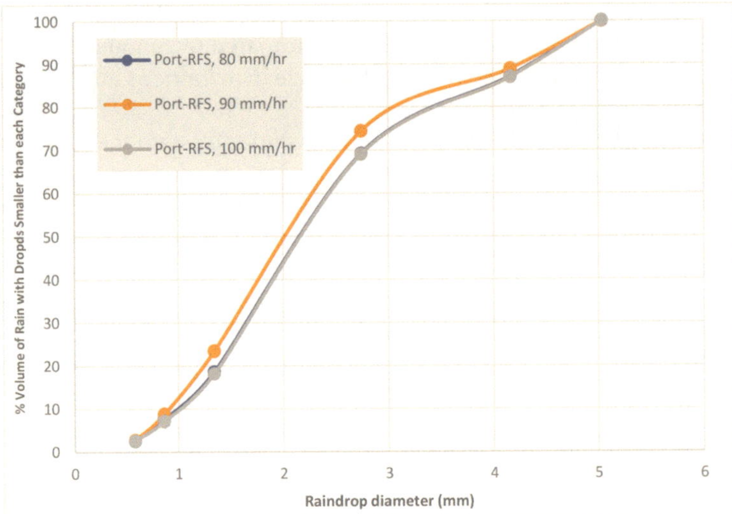

Figure 7.
Raindrop size distribution for the Griffith University portable rainfall simulator (port-RFS) under different simulated rainfall intensities.

drop size distribution pattern observed with the Port-RFS. In **Figure 8**, the relationship between the generated rainfall intensities and the corresponding kinetic energy (KE) per second per flume area is depicted. The measured KE values for rainfall intensities of 80, 90, and 100 mm/h were found to be 1.96, 2.2, and 2.45 Joule/Sec. flume area, respectively. Based on these calibration data, the kinetic energy and drop sizes generated by the Port-RFS were deemed satisfactory for initiating soil erosion in the range of 0.001 mm < D50 < 2.5 mm, which is considered suitable for most soil samples.

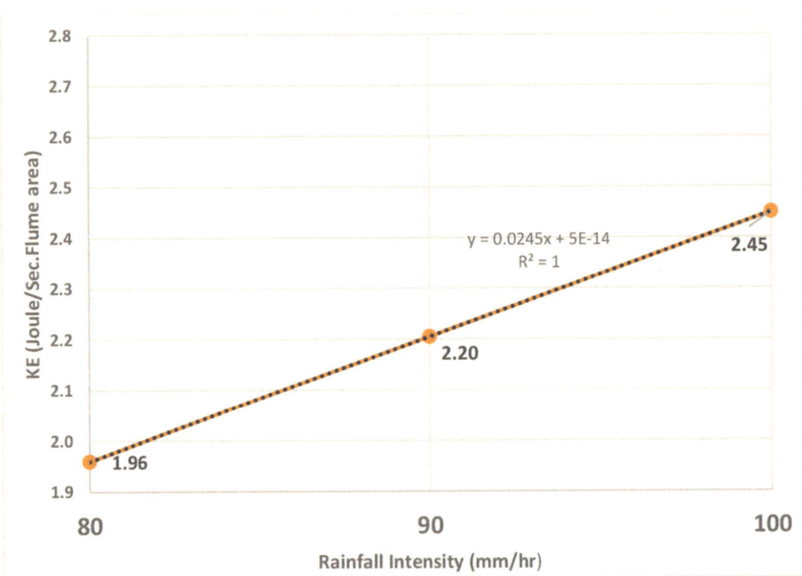

Figure 8.
The relation between rainfall intensity and the kinetic energy (KE) of the simulated rainfall events.

5. Rainfall simulator: the experimental design and data analysis

The experimental design of soil erosion experiments using a rainfall simulator is crucial in obtaining reliable results. The rainfall simulator is designed to simulate natural rainfall events and generate runoff, which can be collected and analyzed for sediment concentration, steady-state runoff rate, and other parameters. The duration and intensity of the rainfall events can be controlled, which allows researchers to investigate the effects of different factors, such as soil type, slope gradient and length, vegetation cover, and land use, on soil erosion.

The experimental design of soil erosion experiments using rainfall simulators should include a detailed description of the soil characteristics, such as texture, organic matter content, and aggregate stability. The location and orientation of the experimental plots, as well as the size and shape of the rainfall simulator nozzle, should also be specified. The experiments should be repeated several times to ensure the reproducibility of the results. The collected runoff should be analyzed for sediment concentration, particle size distribution, and other relevant parameters. Statistical analysis of the data should be performed to determine the significance of the observed differences and to identify the most important factors affecting soil erosion.

Data analysis of soil erosion experiments using rainfall simulators can be challenging due to the complexity of the processes involved. The measured variables are often interdependent, and the relationships between them can be nonlinear. Various statistical methods, such as regression analysis, analysis of variance (ANOVA), and principal component analysis (PCA), can be used to analyze the data. The results should be presented in a clear and concise manner, with appropriate graphs and tables. The conclusions should be based on sound statistical analysis and should be supported by experimental data. The implications of the results for soil management and conservation should also be discussed.

As a practical example of what can be done with soil erosion experiments using rain simulators, Kibet, Saporito [79] presented a protocol along with a video showing how to conduct the experiment; MINErosion 3.4 software [33] also contains a descriptive file that explains the steps for conducting the rainfall simulator experiment and how to calculate rill and interrill soil erodibilities using it.

6. Using the rainfall simulators' results for landform design and assessing the erosional stability of mining rehabilitation sites

Open-cut mining involves a larger disturbance of surface area compared to underground mining. In Australia, explosives are used to blast the deep solid overburden above the mineral or coal seam, which is then mechanically removed using trucks and shovels or draglines. Draglines are the most commonly used methods in Central Queensland open-cut coal mines, which operate at high speeds and result in landscapes consisting of long parallel spoil-piles that are highly saline, dispersive, and erodible. These spoil-piles can be over 50–60 m high and have slopes at an angle of repose of around 75% or 37°. It is legally required for mining organizations to rehabilitate the land by law, and therefore, disturbed open-cut post-mining landscapes must be rehabilitated to an approved post-mining land use. The principal rehabilitation process is shown in **Figure 9**.

Figure 9.
Pictorial representation of the process of rehabilitation of post-mining landscapes on open-cut coal mines is Central Queensland, Australia.

The most expensive part of rehabilitating a mining site is creating a suitable landscape for vegetation growth by reshaping and preparing overburden dumps, which requires costly earthworks. The new landscape must be able to withstand geotechnical failure and surface erosion caused by rainfall and runoff. Steep slopes can lead to severe erosion, causing rehabilitation failures, gully erosion, acid mine drainage (AMD), and salt discharges. Mine sites must evaluate the potential annual erosion rates from rehabilitated areas and report them to regulators, such as the Department of Environment and Science in Queensland, which monitor and enforce rehabilitation compliance. To minimize costs and prevent rehabilitation failures, it is essential to predict the soil erosion rates for the suggested landscape design before its construction. Since the soil and overburden materials used to build the engineered landform have varying degrees of erodibility, accurately measuring their erodibility values is necessary to predict potential erosion rates.

The mining rehabilitation industry has traditionally used predictive equations that rely on soil properties like texture to estimate erodibility values for soil/overburden materials used in constructing engineered landforms. While this approach was easy to use, its unreliability in predicting soil erosion rates led regulators to require actual measurements of soil erodibility values using rainfall simulators in either a field or a laboratory setting. Once these erodibility values have been measured, they are used in combination with more sophisticated erosion/deposition models such as WEPP, MINErosion, and SIBERIA to predict and assess the erosional stability of the proposed landform design.

As an example of using rainfall simulators to estimate and measure soil erodibility values, Sheridan, So [32] conducted a study in which they measured the rill and interrill erodibilities and slope adjustment factors of 34 soil/overburden materials using a portable rainfall simulator tilting flume (measuring 3 m long × 0.8 m wide and with a slope adjustable from 0 to 50%) at the University of Queensland Erosion Processes Laboratory. The materials were exposed to a 100 mm/h rainstorm for 30 min (equivalent to a 1-in-20-year event in Central Queensland) at a 20% slope, followed by slopes of 5, 10, 15, and 30 for 15 min each. At these simulated rainfall intensities, a steady state was quickly reached. Data from rainfall simulation at 10% slope were used to determine interrill erodibility, and data from the overland flow experiments at 20% slope were used to calculate rill erodibility coefficients. Sheridan, So [32] found that soils were generally more erodible than overburdens, as many of the overburdens either contained considerable amounts of rock or had a strong sealing ability. The strongly aggregated high clay soils tended to be the most erodible, followed by the lighter textured sandy loams and loamy sands. Soils or overburdens with 20–30% silt tended to form strong, raindrop impact seals under rainfall and consequently had very low erodibilities. Khalifa [80] expanded the dataset obtained from this research by including information from 93 additional samples (46 soil samples and 47 spoil samples) gathered from six coal mines in Central Queensland. This was done to capture the diversity of spatially distributed samples across each of the selected mining sites. The data collected by Khalifa [80] were used to feed MINErosion 3 & 4 erosion/deposition models for post-mining rehabilitation [81], resulting in the creation of an embedded database file containing the rill and erodibility values for 34 open-cut mine sites in Central Queensland. Additionally, the model is capable of analyzing rainfall simulation data to calculate erodibilities and predict erosion rates on an annual or event-based basis.

Currently, an increasing number of environmental consulting companies operating in the mining sector have constructed and utilized rainfall simulators in a number of environmental rehabilitation projects. Their primary purposes include assessing the erodibility factors of materials intended for constructing the designated land cover and estimating final landform erosion stability. These rainfall simulators have been utilized in several Australian mines, such as the North Parkes Mine in New South

Wales, the Ranger Uranium Mine in the Northern Territory, the Carmichael Coal Mine and Mt. Rawdon gold mine in Queensland, and the Yallourn Coal mine in Victoria.

Furthermore, the utilization of rainfall simulators extends to the assessment of erosion control techniques. By providing a controlled environment, rainfall simulators allow for the testing and evaluation of various measures, such as terracing, vegetation cover, and mulching, to determine their efficacy in reducing soil erosion and improving landform stability [33, 82]. These experiments are valuable in assisting mining operators in selecting the most appropriate and cost-effective erosion control practices tailored to their specific site conditions.

Additionally, rainfall simulators are instrumental in evaluating and monitoring the success of rehabilitation efforts for a specific mine site. After mining operations cease, the restoration of ecosystems and the mitigation of long-term environmental impacts are of utmost importance. Rainfall simulators could be used to facilitate the assessment of rehabilitation effectiveness by comparing the erosional stability of reclaimed landforms with undisturbed reference areas. This evaluation provides invaluable feedback on rehabilitation techniques and offers guidance for future restoration practices.

7. Conclusion

The use of rainfall simulators in assessing the erosional stability of mine site landforms has emerged as a valuable tool for the mining industry. These simulators enable mining operators, researchers, and regulators to measure the material erodibilities, predict erosion rates, estimate the erosional stability, evaluate erosion control measures, and assess the success of land rehabilitation efforts. By providing valuable information on potential erosion hotspots and effective erosion control practices, rainfall simulators contribute to the adoption of responsible mining practices, mitigating environmental impacts, and advancing sustainable land management in the mining sector. When combined with field monitoring and validation, rainfall simulators assume a crucial role in supporting the long-term environmental sustainability of mining operations.

Acknowledgements

The authors express their gratitude to ACARP (the Australian Coal Association Research Project) for their prior support in researching the use of rainfall simulators for rehabilitating mining sites, and to the Australian Rivers Institute-Griffith University for their academic support. Additionally, the corresponding author wishes to acknowledge RGS Environmental for their support and encouragement during his work on this chapter.

A. Appendix

The calculation of Christiansen's uniformity coefficient (CUC) for three different rainfall events simulated by our portable rainfall simulator.

	30 mm/hr. Designed event			60 mm/hr. Designed event			90 mm/hr. Designed event			100 mm/hr. Designed event		
Pan no.	**Measured rainfall intensity (mm/hr)**	**Di**	**|Di-Dm|**	**Measured rainfall intensity (mm/hr)**	**Di**	**|Di-Dm|**	**Measured rainfall intensity (mm/hr)**	**Di**	**|Di-Dm|**	**Measured rainfall intensity (mm/hr)**	**Di**	**|Di-Dm|**
1	28	13.83	1.61	56	28.00	2.29	84	42.17	3.88	91	45.55	6.34
2	37	18.56	3.12	62	31.04	0.75	86	43.18	2.87	88	43.86	8.03
3	34	17.21	1.77	57	28.34	1.95	81	40.49	5.57	88	43.86	8.03
4	38	19.23	3.79	62	31.04	0.75	92	45.88	0.17	101	50.61	1.28
5	31	15.59	0.15	65	32.35	2.06	98	48.82	2.77	106	52.94	1.05
6	35	17.54	2.10	61	30.36	0.07	86	43.18	2.87	101	50.61	1.28
7	31	15.29	0.15	61	30.59	0.29	92	45.88	0.17	112	55.88	3.99
8	38	18.82	3.38	66	32.94	2.65	98	48.82	2.77	118	58.82	6.93
9	29	14.71	0.73	61	30.29	0.00	96	48.24	2.18	118	58.82	6.93
10	27	13.53	1.91	62	31.18	0.88	95	47.65	1.59	118	58.82	6.93
11	29	14.41	1.03	67	33.53	3.24	93	46.47	0.41	118	58.82	6.93
12	21	10.59	4.85	58	28.82	1.47	94	47.06	1.00	106	52.94	1.05
13	22	11.23	4.21	64	31.91	1.62	99	49.65	3.59	106	53.19	1.30
14	35	17.71	2.27	61	30.70	0.41	101	50.61	4.55	103	51.28	0.61
15	35	17.54	2.10	59	29.35	0.94	88	43.86	2.20	88	43.86	8.03
16	26	13.16	2.28	55	27.33	2.97	88	43.86	2.20	101	50.61	1.28
17	28	14.12	1.32	60	30.00	0.29	100	50.00	3.94	106	52.94	1.05
18	30	14.84	0.60	55	27.50	2.80	86	43.18	2.87	101	50.61	1.28
		Dm	15.44		Dm	30.29		Dm	46.06		Dm	51.89
		n	18		n	18		n	18		n	18
		Sum |Di-Dm|	37.37		Sum|Di-Dm|	25.42		Sum |Di-Dm|	45.61		Sum |Di-Dm|	72.35
		CU	86.55		CU	95.34		CU	94.50		CU	92.25

Calculation of Kinetic energy for different rainfall events simulated by portable Rainfall simulator.

$D_r = \sqrt[3]{\left(\frac{6}{\pi}\right) Wm_R}$				(Eq. 1)		$V = \frac{4}{3}\pi\left(\frac{D_r}{2}\right)^3$				(Eq. 4)
Rainfall Intensity	**Number of pellets**	**weights of dry pellets (g)**	**Average weight of dry pallets (mg)**	**m_R**	**D_r (mm)**	**Volume mm^3**	**Density of water 1 mg/mm^3**	**mass of an individual raindrop mg**	**terminal velocity (from the curve below) m/s**	**Kinetic Energy, KE (μJ)**
100	350	2.98	8.520	1.1	3.3588	19.84127	1	19.84127	5.5	300.1
90	387	2.53	6.537	1.1	3.0168	14.37598	1	14.37598	5.5	217.4
60	435	2.29	5.257	1.1	2.7736	11.17245	1	11.17245	5.5	169.0

$$D_r = \sqrt[3]{\left(\frac{6}{\pi}\right)\ Wm_R} \qquad \text{(Eq. 1)}$$

$$V = \frac{4}{3}\pi\left(\frac{D_r}{2}\right)^3 \qquad \text{(Eq. 4)}$$

Rainfall Intensity	Number of pellets	weights of dry pellets (g)	Average weight of dry pallets (mg)	m_R	D_r (mm)	Volume mm^3	Density of water 1 mg/mm^3	mass of an individual raindrop mg	terminal velocity (from the curve below) m/s	Kinetic Energy, KE (μJ)
45	521	1.98	3.797	1.1	2.4670	7.86145	1	7.86145	5.5	118.9
30	567	1.79	3.155	1.1	2.3196	6.534952	1	6.534952	5.5	98.8

Using the flour method to measure the drop size distribution of the Port-RFS.

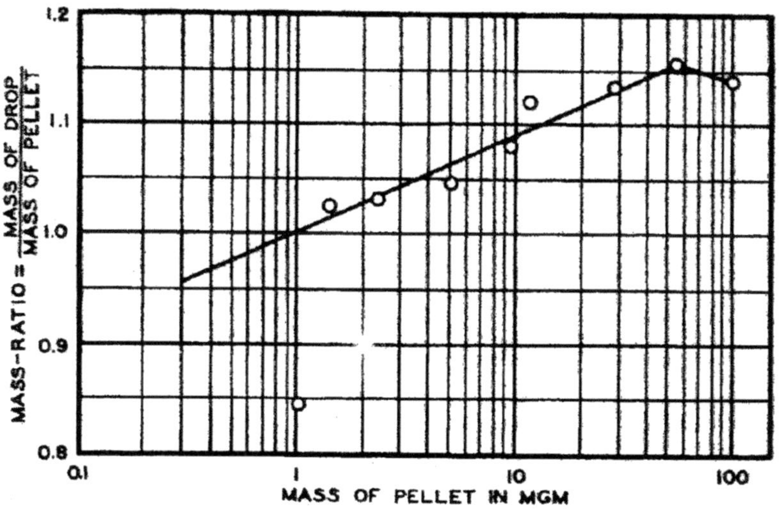

Fig. 2 --Flour-calibration

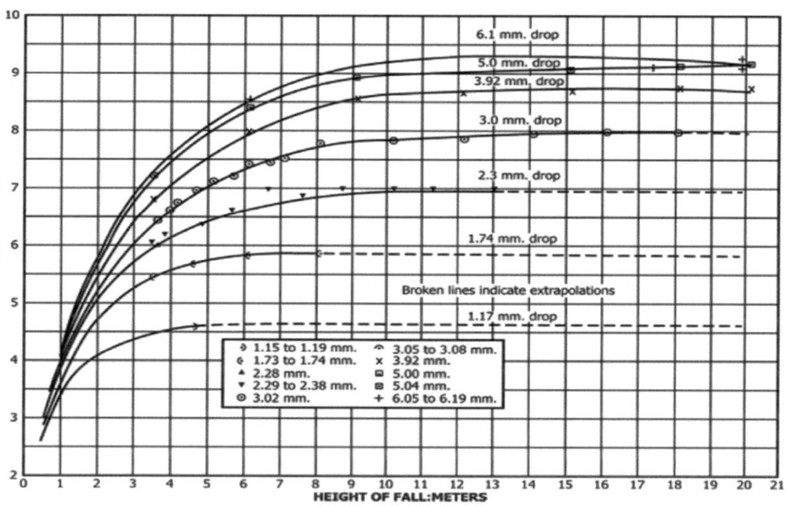

Measuring the water flow rate for different rainfall events simulated by portable RFS.

Time interval	Rainfall intensity	Water pressure (kPa)	Flow rate L/min
2	30	382	0.1
5	40	384	0.3
7	50	386	0.6
10	60	388	0.8
15	70	391	1
20	80	393	1.2
25	90	394	1.5
30	100	395	1.6

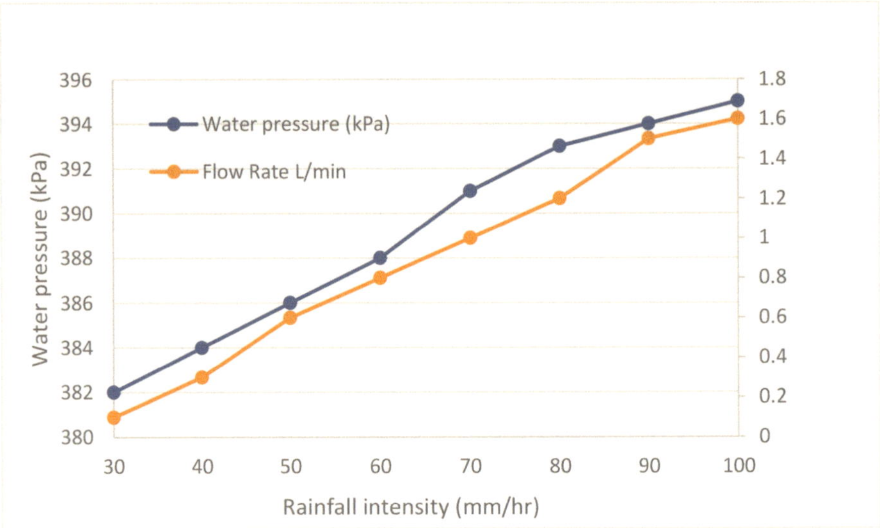

Database of Soil/Spoil Properties (Measured by Khalifa [80]) and Rill/Interrill Erodibilities Assessed with a Portable Rainfall Simulator at 16 Mine Sites in Central Queensland [32, 33, 80].

soil name	interillK	rillK	Infiltration	Coverfactor	EC	pH	CEC	ESP	Wiltingpoint	Fieldcapacity	Organic C	Clay %	Silt%	Sand %	Rock %	BD
Curragh sandstone overburden	2,133,601	4.43	22.61	0.96	0.69	8.89	26.9	26.1	0.12	0.29	0	22.56	25.24	52.2	1.56	1.59
Curragh medium heavy clay soil	6,012,859	11.94	23	0.98	0.75	7.67	60.2	6.5	0.17	0.37	1.6	57.13	17.26	25.62	0.9	1.36
German creek rocky overburden	870,685	5.82	21.98	0.67	1.43	7.57	24.6	10.1	0.08	0.17	0	4.63	28.49	66.88	16.16	1.73
German creek clay loam soil	4,039,506	8.12	21.98	0.9	0.71	7.57	13.2	13.9	0.06	0.17	0.79	22.58	7.92	69.5	4.25	1.59
Goonyella permian overburden	3,048,501	5	15.64	0.97	1.37	8.82	36.5	39.8	0.16	0.31	0	26.85	22.17	50.98	1.31	1.48
Goonyella light clay soil	2,573,589	7.24	19.36	0.96	0.2	6.3	10.1	7.9	0.11	0.18	1.44	44.23	7.23	48.54	1.58	1.58
Gregory overburden	1,629,811	8.06	18.21	0.85	1.05	7.8	16.2	6.5	0.07	0.15	0	9.76	18.16	72.08	6.66	1.79
Gregory clay loam soil	3,969,008	12.72	20.06	0.93	0.07	6.33	13.6	0.5	0.09	0.21	0.78	30.33	5.51	64.16	2.8	1.56
Norwich park rocky overburden	85,664	0.13	70.97	0.44	0.71	8.34	14.6	27.7	0.07	0.12	0	17.22	25.19	57.59	33.14	1.77
Norwich park tertiary overburden	2,874,002	0.29	14.09	0.89	0.48	7.29	15.4	35.4	0.11	0.2	0	31.91	13.04	55.05	4.47	1.61
Norwich park sandy clay loam soil	3,582,186	41.22	12.03	0.97	0.14	6.56	3.5	29.2	0.04	0.11	0.56	15.72	2.71	81.57	1.22	1.73
Oaky Ck shale overburden	796,145	8.93	21.03	0.83	1.6	7.6	30	6.9	0.06	0.16	0	0.72	24.13	75.1	7.34	1.67
Oaky Ck clay loam soil	2,937,891	9.01	10.06	0.98	0.25	7.1	10.2	9.8	0.07	0.17	1.39	22.88	8.47	68.65	0.97	1.61
Wandoan shale overburden	3,123,865	3.09	8.15	0.86	0.85	8.17	25.9	31.3	0.13	0.3	0	26.57	36.62	36.81	6.05	1.41
Wandoan light clay soil	1,861,215	0.83	8.38	0.98	0.11	8.07	23.9	3	0.11	0.23	2.43	33.54	17.94	48.52	0.72	1.48
Callide overburden	1,043,065	5.05	8.94	0.73	0.22	6.68	12.7	5.5	0.08	0.15	0	18.48	11.91	69.61	12.77	1.59
Callide sandy clay loam soil	1,682,373	22.11	8.94	0.91	0.04	5.19	1.4	22.8	0.05	0.13	2.3	19.02	4.31	76.68	3.59	1.59
Blackwater rocky overburden	535,766	17.78	8.86	0.81	0.47	8.62	23	15.9	0.1	0.23	0	20.65	29.26	50.08	8.28	1.65
Blackwater heavy clay soil	5,783,402	21.9	7.58	0.94	0.3	8.69	40.2	10.6	0.21	0.44	2.35	57.65	14.2	28.15	2.45	1.4

soil name	interillK	rillK	Infiltration	Coverfactor	EC	pH	CEC	ESP	Wiltingpoint	Fieldcapacity	Organic C	Clay %	Silt%	Sand %	Rock %	BD
Blair Athol overburden	1,377,148	8.68	4.5	0.81	0.28	7.66	7.6	11.8	0.07	0.16	0	17.72	10.48	71.8	8.55	1.62
Blair Athol sandy loam soil	2,314,952	6.89	2.72	0.86	0.1	7.06	8.6	4	0.06	0.15	1.33	23.69	15.31	61	6.04	1.36
Tarong lithosol soil	1,047,697	6.82	7.06	0.66	0.35	4.99	4.9	14.4	0.09	0.19	3.49	30.06	12.13	57.82	16.77	1.47
Tarong krasnozem soil	760,859	8.06	8.51	0.8	0.09	4.75	2.7	3.3	0.13	0.22	2.63	37.84	6.92	55.24	8.86	1.21
Saraji overburden	4,199,042	2.29	12.74	0.98	0.44	9.44	26.5	28.5	0.13	0.3	0	34	9.08	56.92	0.98	1.5
Saraji sandy loam soil	2,250,000	12.42	5.84	0.97	0.14	8.67	5.8	5.5	0.04	0.13	1.04	14.01	9	76.99	1.26	1.71
Newlands rocky overburden	878,384	7.67	13.01	0.78	0.3	8.16	14.8	3.1	0.07	0.14	0	13.82	18.3	67.88	9.93	1.53
Newlands sandy clay loam soil	3,228,555	2.27	4.66	0.96	0.57	8.93	24.7	36.8	0.1	0.22	0.68	28.58	10.4	61.02	1.49	1.52
Peak downs overburden	1,216,710	1.01	6.22	0.84	1.55	8.55	25.1	52.7	0.12	0.27	0	27.45	42.91	29.64	6.96	1.29
Peak downs loamy sand soil	1,221,103	32.83	6.87	0.94	0.22	7.53	9.5	17.1	0.05	0.14	0.8	20.17	7.03	72.8	2.63	1.44
Moura Medium Clay Soil	2,281,317	7.43	8.03	0.9	0.26	7.81	34.6	9	0.2	0.47	1.67	48.85	15.73	35.42	4.37	1.44
Moura rocky overburden	980,397	1.73	5.46	0.5	0.43	7.88	25.1	19.2	0.14	0.29	0	24.13	25.38	50.49	28.01	1.48
Moura permian overburden	3,297,501	4.81	0.14	0.88	0.75	8.57	32.6	31.3	0.12	0.26	0	31.06	16.66	52.28	4.93	1.18
Tarong sandstone overburden	2,500,000	6	15	0.98			2.2	3.26	0.12	0.26	0	12	16	72	8	
Kidston	260,394	0.61	5	0.5	0.06	8.5	17.4	8.04								

Author details

Ashraf M. Khalifa[1,2]* and Hwat Bing So[1]

1 Australian Rivers Institute, Griffith University, Nathan, Queensland, Australia

2 SLR Consulting Australia Pty Ltd., Brisbane QLD, Australia

*Address all correspondence to: a.khalifaaly@griffith.edu.au

References

[1] Browning GM, Parish CL, Glass J. A method for determining the use and limitations of rotation and conservation practices in the control of soil erosion in Iowa1. Agronomy Journal. 1947;**39**(1): 65-73

[2] Welsh D et al. Coal mines on target with environmental planning. Queensland Government Mining Journal;(Australia). 1994;**95**(1107):19-22

[3] Williams DJ. Assessment of embankment parameters. Slop Stability in Surface Mining. In: Hustrulid WA, McCarter MC, van Zyl DJA. Littleton, Colorado, USA: Society of Mining, Metallurgy and Exploration; 2001. pp. 275-284

[4] Lu H et al. Predicting sheetwash and rill erosion over the Australian continent. Australian Journal of Soil Research. 2003;**41**(6):1037-1062

[5] Hamed Y et al. Comparison between rainfall simulator erosion and observed reservoir sedimentation in an erosion-sensitive semiarid catchment. Catena. 2002;**50**(1):1-16

[6] Iserloh T et al. A small portable rainfall simulator for reproducible experiments on soil erosion. Soil and Tillage Research. 2012;**124**:131-137

[7] Loch RJ et al. A multi-purpose rainfall simulator for field infiltration and erosion studies. Australian Journal of Soil Research. 2001;**39**:599

[8] Macedo MS, , et al. A modified portable rainfall simulator for soil erosion assessment under different rainfall patterns. Journal of Hydrology. 2021;**596**:126052

[9] Mhaske SN, Pathak K, Basak A. A comprehensive design of rainfall simulator for the assessment of soil erosion in the laboratory. Catena. 2019; **172**:408-420

[10] Vahabi J, Nikkami D. Assessing dominant factors affecting soil erosion using a portable rainfall simulator. International Journal of Sediment Research. 2008;**23**(4):376-386

[11] Vergni L, Todisco F, Vinci A. Setup and calibration of the rainfall simulator of the masse experimental station for soil erosion studies. Catena. 2018;**167**: 448-455

[12] Wu B et al. Modelling sheet erosion on steep slopes of clay loess soil using a rainfall simulator. Biosystems Engineering. 2022;**216**:1-12

[13] Shao Q, Baumgartl T. Estimating input parameters for four infiltration models from basic soil, vegetation, and rainfall properties. Soil Science Society of America Journal. 2014;**78**(5): 1507-1521

[14] Wierda A, Veen AWL. A rainfall simulator study of infiltration into arable soils. Agricultural Water Management. 1992;**21**(1):119-135

[15] Boulange J et al. Portable rainfall simulator for plot-scale investigation of rainfall-runoff, and transport of sediment and pollutants. International Journal of Sediment Research. 2019; **34**(1):38-47

[16] Mamoon AA et al. First flush analysis using a rainfall simulator on a micro catchment in an arid climate. Science of the Total Environment. 2019; **693**:133552

[17] Regmi TP, Thompson AL. Rainfall simulator for laboratory studies. Applied

Engineering in Agriculture. 2000;**16**(6): 641-647

[18] Cecconi M, Napoli P, Pane V. Effects of soil vegetation on shallow slope instability. Environmental Geotechnics. 2015;**2**(3):130-136

[19] Egeli I, Pulat HF. Mechanism and modelling of shallow soil slope stability during high intensity and short duration rainfall. Scientia Iranica. 2011;**18**(6): 1179-1187

[20] Huang C-C, Yuin S-C. Experimental investigation of rainfall criteria for shallow slope failures. Geomorphology. 2010;**120**(3):326-338

[21] Lora M, Camporese M, Salandin P. Design and performance of a nozzle-type rainfall simulator for landslide triggering experiments. Catena. 2016; **140**:77-89

[22] Mendes TA et al. Development of a rainfall and runoff simulator for performing hydrological and geotechnical tests. Sustainability. 2021; **13**(6):3060

[23] Montrasio L, Schilirò L, Terrone A. Physical and numerical modelling of shallow landslides. Landslides. 2016; **13**(5):873-883

[24] Aksoy H et al. A rainfall simulator for laboratory-scale assessment of rainfall-runoff-sediment transport processes over a two-dimensional flume. Catena. 2012;**98**:63-72

[25] Nichols M, Sexton H. A method of studying soil erosion. Agricultural Engineering. 1932;**13**(4):101-103

[26] Kavka P et al. Chapter 17-rainfall simulation experiments as a tool for process research in soil science, hydrology, and geomorphology. In: Rodrigo-Comino J, editor. Precipitation. Elsevier; 2021. pp. 395-418. DOI: 10.1016/B978-0-12-822699-5.00015-X

[27] Rončević V et al. Dripping rainfall simulators for soil research—Performance review. Water. 2023;**15**(7): 1314

[28] Neumann M et al. Effect of plot size and precipitation magnitudes on the activation of soil erosion processes using simulated rainfall experiments in vineyards. Frontiers in Environmental Science. 2022;**10**. DOI: 10.3389/fenvs.2022.949774

[29] Danáčová M, Valent P, Výleta R. Evaluation of surface runoff generation processes using a rainfall simulator: A small scale laboratory experiment. In: IOP Conference Series: Earth and Environmental Science. 2017. DOI: 10.1088/1755-1315/95/2/022016

[30] Iserloh T, Ries JB, Seeger M. Requirements for Future Development of Small Scale Rainfall Simulators 2013. p. EGU2013-10739. Available from: https://ui.adsabs.harvard.edu/abs/2013EGUGA..1510739I

[31] Dunkerley D. The case for increased validation of rainfall simulation as a tool for researching runoff, soil erosion, and related processes. Catena. 2021;**202**: 105283

[32] Sheridan GJ et al. Use of laboratory-scale rill and interill erodibility measurements for the prediction of hillslope-scale erosion on rehabilitated coal mine soils and overburdens. Australian Journal of Soil Research. 2000;**38**(2):285-297

[33] So HB et al. MINErosion 3: Using measurements on a tilting flume-rainfall simulator facility to predict erosion rates

from post-mining landscapes in Central Queensland, Australia. PLoS One. 2018; **13**(3):e0194230

[34] Loch RJ, Silburn DM, Freebairn DM. Evaluation of the CREAMS model. II. Use of rainulator data to derive soil erodibility parameters and prediction of eield soil losses using derived parameters. Australian Journal of Soil Research. 1989;**27**:563-576

[35] Silburn DM, Connolly RD. Distributed parameter hydrology model (ANSWERS) applied to a range of catchment scales using rainfall simulator data. I application to rainfall simulator plots. Journal of Hydrology. 1995;**172**:87-104

[36] Silburn DM, Loch RJ. Present capabilities and constraints in modelling soil erosion. In: Hamilton GJ, Howes KM, Attwater R, editors. Erosion Productivity and Erosion Prediction Workshop. Western Australia: 5th Australian Soil Conservation Conference; 1990

[37] Abudi I, Carmi G, Berliner P. Rainfall simulator for field runoff studies. Journal of Hydrology. 2012; **454–455**:76-81

[38] Alves Sobrinho T, Gómez-Macpherson H, Gómez JA. A portable integrated rainfall and overland flow simulator. Soil Use and Management. 2008;**24**(2):163-170

[39] Au-Kibet LC et al. A protocol for conducting rainfall simulation to study soil runoff. JoVE. 2014;**86**: e51664

[40] Blanquies J et al. The design and construction of a rainfall simulator. In: International Erosion Control Association (IECA). Las Vegas, Nevada: 34th Annual Conference and Expo; February, 2003; 2003. pp. 24-28.

[41] Bowyer-Bower TAS, Burt TP. Rainfall simulators for investigating soil response to rainfall. Soil Technology. 1989;**2**(1):1-16

[42] Claassens AS, van der Watt HvH. An inexpensive, portable rain simulator: Construction and test data. South African Journal of Plant and Soil. 1993; **10**(1):6-11

[43] Moore ID, Hirschi MC, Barfield BJ. Kentucky rainfall simulator. Transactions of the ASAE. 1983;**26**(4): 1085-1089

[44] Esteves M et al. The 'EMIRE' large rainfall simulator: Design and field testing. Earth Surface Processes and Landforms. 2000;**25**:681-690

[45] Grierson IT, Oades JM. A rainfall simulator for field studies of run-off and soil erosion. Journal of Agricultural Engineering Research. 1977;**22**(1):37-44

[46] Horne MA. Design and construction of a rainfall simulator for large-scale testing of erosion control practices and products. In: Graduate Faculty of Auburn University. Vol. 166. Auburn, Alabama: Auburn University; 2017

[47] Kiwan ME, Al-Wagdany AS. Design and construction of a rainfall simulator. MISR Journal of Agricultural Engineering. 2009;**26**(2):714-725

[48] Meyer LD, McCune DL. Rainfall simulator for runoff plots. Agricultural Engineering. 1958;**39**(10):644-648

[49] Ngezahayo E, Michael PNB, Ghataora GS. Rainfall Induced Erosion of Soils Used in Earth Roads. Les Ulis: EDP Sciences; 2019

[50] Ricks MD et al. Design of a pressurized rainfall simulator for evaluating performance of erosion

control practices. Water. 2019;**11**(11): 2386

[51] Sangüesa C et al. A rainfall simulator for the in situ study of superficial runoff and soil erosion. Chilean journal of agricultural research. 2010;**70**:178-182. Available from: http://www.scielo.cl/scielo.php?script=sci_arttext&pid=S0718-58392010000100019&nrm=iso

[52] Szabó B et al. Soil Erodibility Variability in Laboratory and Field Rainfall Simulations. 2017. p. 13725

[53] Živanović N, Rončević V, Spasić M, Ćorluka S, Polovina S. Construction and calibration of a portable rain simulator designed for the in situ research of soil resistance to erosion. Soil and Water Research. 2022;**17**:158-169

[54] Lassu T et al. The Wageningen rainfall simulator: Set-up and calibration of an indoor nozzle-type rainfall simulator for soil erosion studies. Land Degradation & Development. 2015; **26**(6):604-612

[55] Calibration of DISTFW parameters for QDPI rainfall simulator : German Creek Mine / Y. Gyasi-Agyei ... [et al.]. Research Report (University of Newcastle (NSW.). Department of Civil Engineering and Surveying) 120.11.1995., ed. Y. Gyasi-Agyei, E. University of Newcastle. Department of Civil, and Surveying. Callaghan, NSW: Department of Civil Engineering and Surveying, University of Newcastle; 1995

[56] Bubenzer GD. Rainfall characteristics important for simulation. In: Proceedings of the Rainfall Simulator Workshop, D.o.a.S.a. Education, Editor. Tucson Arizona: Administration Agricultural Reviews and Manuals; 1979. pp. 22-34

[57] Christiansen JE. Hydraulics of sprinkling Systems for Irrigation. Transactions of the American Society of Civil Engineers. 1942;**107**(1):221-239

[58] Sousa S, Mendes T, Siqueira E. Development and calibration of a rainfall simulator for hydrological studies. RBRH Scientific/Technical Article. 2017;**22**

[59] Luk S-h, Abrahams AD, Parsons AJ. Sediment sources and sediment transport by rill flow and interrill flow on a semi-arid piedmont slope, southern Arizona. Catena. 1993;**20**(1): 93-111

[60] Serio MA, Carollo FG, Ferro V. Raindrop size distribution and terminal velocity for rainfall erosivity studies. A review. Journal of Hydrology. 2019;**576**: 210-228

[61] Hall MJ. Use of the stain method in determining the drop-size distributions of coarse liquid sprays. Transactions of the ASAE. 1970;**13**(1):33-0037

[62] Bentley WA. Studies of raindrops and raindrop phenomena. Monthly Weather Review. 1904;**32**(10):450-456

[63] Eigel JD, Moore ID. A simplified technique for measuring raindrop size and distribution. Transactions of the ASAE. 1983;**26**(4):1079-1084

[64] Laws JO. Measurements of the fall-velocity of water -drops and raindrops. Eos, Transactions American Geophysical Union. 1941;**22**(3):709-721

[65] Joss J, Gori EG. Shapes of raindrop size distributions. Journal of Applied Meteorology (1962–1982). 1978: 1054-1061

[66] Kinnell P. Rainfall intensity kinetic energy relationship for soil loss prediction. Soil Science Society of America Journal. 1984;**27**:1836-1840

[67] Hoffman F. Applications of Droplet Photography. Springfield, MA: Calfran Industries; 1977

[68] Jayawardena A, Rezaur R. Measuring drop size distribution and kinetic energy of rainfall using a force transducer. Hydrological Processes. 2000;**14**(1):37-49

[69] Bringi VN et al. Raindrop size distribution in different climatic regimes from disdrometer and dual-polarized radar analysis. Journal of the Atmospheric Sciences. 2003;**60**(2): 354-365

[70] Gorgucci E et al. Rainfall estimation from polarimetric radar measurements: Composite algorithms immune to variability in raindrop shape-size relation. Journal of Atmospheric and Oceanic Technology. 2001;**18**(11): 1773-1786

[71] Salles C, Poesen J. Performance of an optical spectro pluviometer in measuring basic rain erosivity characteristics. Journal of Hydrology. 1999;**218**(3–4): 142-156

[72] Fernández-Raga M et al. Rain events on 22 October 2006 in León (Spain): Drop size spectra. Atmospheric Research. 2009;**93**(1):619-635

[73] Fraile R et al. Error in the sampling area of an optical disdrometer: Consequences in computing rain variables. The Scientific World Journal. 2013;**2013**:369450

[74] Fernández-Raga M et al. Splash erosion: A review with unanswered questions. Earth-Science Reviews. 2017; **171**:463-477

[75] Rodrigo-Comino J et al. Role of rock fragment cover on runoff generation and sediment yield in tilled vineyards. European Journal of Soil Science. 2017; **68**(6):864-872

[76] Laws JO, Parson DA. The relation of raindrop-size to intensity. Transactions of the American Geophysical Union, Papers Hydrology. 1943:453-460

[77] ASTM. Standard Test Method for Determination of Rolled Erosion Control Product Performance in Protecting Hillslopes from Rainfall-Induced Erosion. West Conshohocken, PA: American Standards for Testing Materials; 2015

[78] Salles C, Poesen J, Govers G. Statistical and physical analysis of soil detachment by raindrop impact: Rain erosivity indices and threshold energy. Water Resources Research. 2000;**36**(9):2721-2729. DOI: 10.1029/2000WR900024

[79] Kibet LC et al. A protocol for conducting rainfall simulation to study soil runoff. Journal of Visualized Experiments. 2014;(86)

[80] Khalifa AM. MINErosion 4: A user-friendly catchment/landscape erosion prediction model for post mining sites in Central Queensland. In: Griffith School of Engineering. Brisbane, Queensland, Australia: Griffith University; 2010

[81] Khalifa AM et al. MINErosion 4: Using Measurements from a Tilting Flume-Rainfall Simulator Facility to Predict Erosion Rates from Post-Mining Catchments/Landscapes in Central Queensland, Australia. International Soil and Water Conservation Research; 2023

[82] Loch RJ. Using rainfall simulation to guide planning and management of rehabilitated areas: Part I. experimental methods and results from a study at the Northparkes mine, Australia. Land Degradation & Development. 2000; **11**(3):221-240